AF326863

Current Topics in Microbiology and Immunology

Volume 428

The review series *Current Topics in Microbiology and Immunology* provides a synthesis of the latest research findings in the areas of molecular immunology, bacteriology and virology. Each timely volume contains a wealth of information on the featured subject. This review series is designed to provide access to up-to-date, often previously unpublished information.

2019 Impact Factor: 3.095., 5-Year Impact Factor: 3.895

2019 Eigenfaktor Score: 0.00081, Article Influence Score: 1.363

2019 Cite Score: 6.0, SNIP: 1.023, h5-Index: 43

More information about this series at http://www.springer.com/series/82

Lars Hangartner · Dennis R. Burton
Editors

Vaccination Strategies Against Highly Variable Pathogens

Responsible Series Editor: Peter K. Vogt

 Springer

Editors
Lars Hangartner
Department of Immunology
and Microbiology
The Scripps Research Institute
La Jolla, CA, USA

Dennis R. Burton
Department of Immunology
and Microbiology
The Scripps Research Institute
La Jolla, CA, USA

ISSN 0070-217X   ISSN 2196-9965   (electronic)
Current Topics in Microbiology and Immunology
ISBN 978-3-030-58003-2   ISBN 978-3-030-58004-9   (eBook)
https://doi.org/10.1007/978-3-030-58004-9

This Springer imprint is published by the registered company Springer Nature Switzerland AG
The registered company address is: Gewerbestrasse 11, 6330 Cham, Switzerland

Preface

Recent emergence of various viral diseases clearly demonstrates that vaccine development is more important than ever. Classical vaccine approaches involving administration of inactivated pathogens, or live pathogens that have been attenuated, have been very successful in curbing or even eradicating viral diseases that show limited antigenic variability. However, these classical approaches have been unsuccessful when employed to elicit protective immunity against highly antigenically variable pathogens, such as HIV or influenza virus, or pathogens that exist in multiple serotypes. In this book, we illustrate how modern vaccine approaches aim to cope with this variability, as research efforts have focused on identification of protective antibody epitopes that are shared between different serotypes, or that are less variable because they are directly involved in crucial infection processes for the pathogen.

Great strives have been made to overcome the many obstacles in directing the immune response toward shared or conserved epitopes lays, and protein engineering has enabled expression of various surface antigens such that they accurately represent the antigenic structure of the pathogens. For viral antigens involved in cell entry, this has proven to be particularly difficult as these proteins are typically present in a 'spring-loaded' metastable state on infectious virions. Following triggering, the enthalpy contained in this metastable state is used to power the conformational changes that ultimately lead to entry into the host cell and that leaves the protein in this energetically preferred, post-fusion state. It is therefore not surprising that when recombinantly expressed as soluble proteins, many viral surface antigens end up in the post-fusion configuration. For HIV envelope, it took 30 years of research to identify stabilizing mutations that successfully lock the protein in the pre-fusion state and that allowed expression of soluble trimers that accurately mimic the antigenic structure of the virion. However, despite these successes, it has become apparent that shared epitopes are typically very poorly immunogenic and will require intensive research to develop vaccination strategies, and adjuvants, to allow for reproducible induction of antibodies against conserved epitopes.

While all these efforts are yet to yield vaccines in the clinic, we have gained a better understanding of antigen design, immunodominance, analysis of antibody responses and vaccine formulations, and the toolset developed for highly antigenically variable viruses has become an essential weapon in our fight against emerging viruses where time is of the essence.

La Jolla, USA

Lars Hangartner
Dennis R. Burton

Contents

Successful Vaccines

Ian J. Amanna and Mark K. Slifka

Contents

Abstract Vaccines are considered one of the most important advances in modern medicine and have greatly improved our quality of life by reducing or eliminating many serious infectious diseases. Successful vaccines have been developed against many of the most common human pathogens, and this success has not been dependent upon any one specific class of vaccine since subunit vaccines, non-replicating whole-virus or whole-bacteria vaccines, and attenuated live vaccines have all been

I. J. Amanna
Najít Technologies, Inc, Beaverton, OR 97006, USA
e-mail: iamanna@najittech.com

M. K. Slifka (✉)
Oregon National Primate Research Center, Oregon Health
& Science University, Beaverton, OR 97006, USA
e-mail: slifkam@ohsu.edu

Current Topics in Microbiology and Immunology (2020) 428: 1–30
https://doi.org/10.1007/82_2018_102

Published Online: 26 July 2018

effective for particular vaccine targets. After completing the initial immunization series, one common aspect of successful vaccines is that they induce long-term protective immunity. In contrast, several partially successful vaccines appear to induce protection that is relatively short-lived and it is likely that long-term protective immunity will be critical for making effective vaccines against our most challenging diseases such as AIDS and malaria.

1   Introduction

The effect of vaccines on public health is truly remarkable. One study examining the impact of childhood vaccination on the 2001 US birth cohort found that vaccines prevented 33,000 deaths and 14 million cases of disease (Zhou et al. 2005). Among 73 nations supported by the GAVI Alliance, mathematical models project that vaccines will prevent 23.3 million deaths from 2011 to 2020 compared to what would have occurred if there were no vaccines available (Lee et al. 2013). Vaccines have been developed against a wide assortment of human pathogens (Table 1). There are vaccines against bacterial toxins (e.g., tetanus and diphtheria toxins), acute viral pathogens (e.g., measles, mumps, rubella), latent or chronic viral pathogens (e.g., varicella zoster virus [VZV] and human papilloma virus [HPV], respectively), respiratory pathogens (e.g., influenza, *Bordetella pertussis*), and enteric pathogens (e.g., poliovirus, *Salmonella typhi*). Most licensed vaccines can

Table 1 Examples of licensed vaccines

Live, attenuated vaccines
Adenovirus types 4 and 7
Dengue virus[a]
Influenza virus[b]
Measles virus
Mumps virus
Mycobacterium tuberculosis
Poliovirus[c]
Rotavirus
Rubella virus
Salmonella typhi (typhoid fever)
Varicella zoster virus (chicken pox and shingles)
Variola virus (smallpox)
Yellow fever virus
Non-replicating microbial particle vaccines
Bordetella pertussis (whooping cough)[d]
Hepatitis A virus

(continued)

Table 1 (continued)

Non-replicating whole particle vaccines
Hepatitis B virus
Human papillomavirus
Influenza virus[e]
Japanese encephalitis virus
Neisseria meningitidis serogroup B (bacterial meningitis)[f]
Poliovirus
Rabies virus
Tick-borne encephalitis virus[g]
Yersinia pestis (bubonic plague)[h]
Subunit vaccines
Bacillus anthracis (anthrax)
Bordetella pertussis (whooping cough)
Diphtheria toxin
Haemophilus influenzae type b
Influenza virus
Neisseria meningitidis serogroups A, C, Y, W-135, and B (bacterial meningitis)[f]
Salmonella typhi (typhoid fever)
Streptococcus pneumoniae (pneumococcal disease)
Tetanus toxin
Varicella zoster virus (shingles)

[a]As of October 11, 2017, Dengvaxia® was registered in 19 countries (Sanofi Pasteur 2017). In December 2017, the Philippines suspended sale of the vaccine due to safety concerns (Bacungan et al. 2018)

[b]FluMist was not recommended for the 2017–2018 influenza season by the ACIP due to concerns regarding its effectiveness (Grohskopf et al. 2017)

[c]In the USA, OPV was replaced with IPV by 2000 (Advisory Committee on Immunization Practices 2000)

[d]The ACIP recommended the full replacement of whole-cell pertussis (wP) with the acellular pertussis (aP) vaccine by 1997 in the USA (Advisory Committee on Immunization Practices 1997), but the wP vaccine continues to be used routinely in other countries

[e]A whole-virion inactivated H5N1 vaccine (Vepacel®) is licensed for use in the EU (Plosker 2012)

[f]The Bexsero® meningococcal group B vaccine uses a combination of outer membrane vesicle particles and individual recombinant proteins (Novartis Vaccines and Diagnostics 2015)

[g]A tick-borne encephalitis virus vaccine is not currently licensed for the USA, but is used in multiple EU countries (Heinz et al. 2013)

[h]Manufacture of the plague vaccine in the USA was discontinued in 1999 (Williamson and Oyston 2013)

be categorized as (a) live, attenuated vaccines, (b) non-replicating whole-particle vaccines (including virus-like particles or VLPs), and (c) subunit vaccines. There are similar numbers of licensed vaccines in each of these categories (Table 1), and this illustrates the point that no single vaccine approach is superior to another, but instead the approach varies according to each individual pathogen and the feasibility, safety, and efficacy required to develop a successful vaccine to that particular pathogen. Interestingly, influenza represents the only example in which there is at least one licensed vaccine listed in each of these three categories. However, while still licensed the live-attenuated influenza vaccine (FluMist) was removed from the list of recommended vaccines for the 2017–2018 season due to concerns about its effectiveness against influenza A(H1N1)pdm09 (Grohskopf et al. 2017).

2   Successes and Failures of Live, Attenuated Vaccines

There are currently 13 diseases for which live, attenuated vaccines have been developed and licensed for commercial use (Table 1). Development of successful vaccines is typically easier when there is (a) a single serotype, (b) the pathogen is antigenically stable, (c) diagnostic tests/clinical criteria for measuring disease burden are available, and (d) the vaccine is both safe and effective. Smallpox vaccination with vaccinia virus is the most famous example of a highly effective vaccine, and at the time when people were faced with smallpox outbreaks, this vaccine was associated with each of these characteristics that led to the implementation of a successful vaccine. The smallpox vaccine is not only the first vaccine ever developed (Jenner 1798, 1799, 1800), but since variola virus (the causative agent of smallpox) no longer exists in nature, this also represents the first example in which a vaccine was purposefully used to drive a species to extinction. Some of the best examples of successful live, attenuated vaccines include those developed against measles, mumps, and rubella (MMR). In contrast, other live, attenuated vaccines such as the oral polio vaccine (OPV) and the first licensed vaccine against rotavirus (RotaShield[®]) serve as important reminders that vaccines must be safe if they are going to be used successfully for routine vaccination.

2.1   *Measles, Mumps, Rubella*

One of the most common live, attenuated vaccines in use today is the MMR vaccine. Although originally designed as three separate vaccines, attenuated strains of measles, mumps, and rubella were successfully formulated into a single combination vaccine in the USA in 1971 (Strebel et al. 2013) and it continues to be used to provide protective immunity against pathogens responsible for what were once considered common childhood diseases. Similar to smallpox vaccination, they were developed against viruses that are antigenically stable and have only a single serotype. Even when combined with live, attenuated VZV to provide a quadrivalent vaccine, seroconversion rates remain high (measles: 97.1%, mumps: 96.0%, rubella: 98.8%, and varicella: 93.5%, respectively) (Lieberman et al. 2006). Immunity, particularly against mumps, may wane significantly over time (Cardemil et al. 2017), and the Advisory Committee on Immunization Practices (ACIP) recommends a third dose be administered to close contacts during outbreaks

(Marin et al. 2018). Other vaccines against antigenically variable viruses (e.g., influenza) or pathogens with multiple serotypes (e.g., HPV) indicate that as vaccine technologies have improved and evolved, there are opportunities to combat more complicated and challenging pathogens through effective immunization programs. Although nearly all vaccines require at least one booster dose in order to maintain long-term protective immunity (Slifka and Amanna 2014), many of our most successful vaccines have achieved the appropriate balance of both safety and efficacy.

2.2 Rotavirus and Poliovirus

There are many aspects to a vaccine that are required for them to be considered successful, and one of the most important parameters for routine vaccines is patient safety. When vaccines demonstrate safety and efficacy during Phase II/III trials, then they have the opportunity to be licensed for commercial production while continuing to be monitored post-licensure (i.e., Phase IV). The RotaShield vaccine developed against rotavirus represents a vaccine that failed post-licensure due to an unacceptably high rate of reactogenicity in which serious adverse events (e.g., intussusception) reached 90–214 cases per million administered doses (Kramarz et al. 2001; Murphy et al. 2001). Following licensure in 1998, it was subsequently pulled from the market by the end of 1999. Live, attenuated rotavirus vaccines with improved safety profiles (Rotarix® and RotaTeq®) have now entered the market to address this important clinical need. Similarly, live OPV provided good immunogenicity and protective efficacy, but also caused vaccine-associated paralytic poliomyelitis (VAPP) at a rate of 0.4–0.5 cases per million administered doses (Hao et al. 2008; Prevots et al. 1994). Because of its widespread use in the USA, this resulted in an average of nine children each year from 1961 to 1989 who suffered paralysis due to OPV (Alexander et al. 2004). Since polio was no longer endemic in the USA, the OPV vaccine was causing more cases of paralysis than the wild-type polioviruses that it was designed to prevent. In 2000, the OPV vaccine regimen was abandoned in favor of the safer inactivated polio vaccine (IPV) series and VAPP has been eliminated (Alexander et al. 2004). Likewise, the complete worldwide replacement of OPV in favor of IPV is a key objective of the Global Polio Eradication Initiative, with the goal of completing this switch by 2020 (Garon and Patel 2017).

3 Successes and Failures of Non-replicating/Inactivated Vaccines

Successful non-replicating/inactivated vaccines have been licensed for 11 different pathogens and include both viruses and bacteria (Table 1). As described below, the hepatitis A virus (HAV) vaccine represents a striking success for whole-virion inactivated vaccines, with exceptional efficacy and long-term antibody persistence demonstrated for up to 20 years post-vaccination (Theeten et al. 2015). The HPV vaccine utilizes advanced VLP technology and is able to protect against chronic viral infection, prevent HPV-associated cancers, and potentially induce lifelong immunity with as few as two doses or even a single immunization (Kreimer et al. 2015; Safaeian et al. 2013). In contrast to these success stories, unsuccessful attempts to develop formaldehyde-inactivated measles and respiratory syncytial virus (RSV) vaccines in the 1960s indicate that an inactivated whole-virus vaccine approach is not foolproof and several factors may have contributed to those failures (Polack 2007).

3.1 Hepatitis A

Infection with HAV was once a leading cause of viral hepatitis in the USA, with 20,000–60,000 cases reported annually and fatality rates as high as 1.8% in adults >50 years of age (Centers for Disease Control and Prevention 2015c). Vaccine development against HAV benefited from several key factors including early studies demonstrating the efficacy of passive immunotherapy (Gellis et al. 1945; Havens and Paul 1945; Hsia et al. 1954; Stokes and Neefe 1945), the discovery of cell culture conditions for the propagation of HAV (Provost and Hilleman 1979), and passive and active immunization studies in appropriate animal models including chimpanzees and marmosets (Provost et al. 1986; Purcell et al. 1992). These critical elements supported the eventual development and licensure of two inactivated whole-virion, alum-adjuvanted vaccines (Havrix$^{®}$ and Vaqta$^{®}$) in the USA, and since their implementation in the 1990s, the rates of disease have declined by >90% (Centers for Disease Control and Prevention 2015c). Efficacy for both vaccines is high, with estimates ranging from 94 to 100% (Innis et al. 1994; Werzberger et al. 1992). HAV-specific antibodies have been identified as the key correlate of protective immunity following vaccination, and although an absolute lower limit has not been defined, titers >10–20 milli-international units per milliliter (mIU/mL) are generally considered protective (Centers for Disease Control and Prevention 1996).

One of the most impressive aspects of the inactivated HAV vaccines has been the durability of vaccine-induced antibody responses. Preliminary studies examining HAV-specific antibody titers demonstrated a rapid drop shortly after vaccination, but also showed a second, slower rate of decay, with protective immunity estimated to be maintained for 20–30 years following vaccination (Van Damme et al. 1994). Longer-term studies have confirmed and extended these initial results (Rendi-Wagner et al. 2007; Theeten et al. 2015; Wiedermann et al. 1997). In one notable study, antibody titers in subjects who received two doses of inactivated HAV vaccine were followed annually for up to 20 years post-vaccination (Theeten et al. 2015). Based on these results, it was estimated that at 40 years post-vaccination, $\geq$ 90% of vaccinees would maintain antibodies above the putative protective threshold. This may actually be an underestimate since, as noted by the authors, antibody decline appeared to level off over time with minimal changes to average group antibody titers from 11 to 20 years post-vaccination (Theeten et al. 2015). Remarkably, this data indicates that two doses of an alum-adjuvanted inactivated vaccine may be sufficient to induce lifelong immunity for most vaccine recipients.

3.2 Human Papilloma Virus

Exposure to HPV, typically transmitted at mucosal surfaces through sexual contact, can lead to persistent infection and a range of cancers (Centers for Disease Control and Prevention 2015d). Cervical cancer, caused almost exclusively by HPV, still bears a substantial burden on the USA, with 4074 deaths reported in 2012 alone [the most recent year with available data (Centers for Disease Control and Prevention 2015a)]. Although more than 120 HPV types have been identified, a smaller subset of approximately 13–14 types is considered high risk for developing cancer (Centers for Disease Control and Prevention 2015d). Successful development of HPV vaccines was precipitated by the discovery that the L1 capsid protein of the virus could self-assemble into VLPs and induce high-titer neutralizing antibodies (Kirnbauer et al. 1992). Papillomaviruses are species-restricted, and this constrains the ability to assess HPV vaccine candidates directly in animals. However, several species-specific models (including mucosal challenge routes) were tested using L1 VLP candidates (Schiller and Lowy 1996). These animal studies demonstrated several key findings including: (a) VLPs could protect prophylactically against disease but had no impact on established infections, (b) denatured VLPs were unable to elicit protective responses, suggesting that the three-dimensional conformation of the VLP was important to eliciting appropriate immunity, and (c) passively transferred antibodies from immunized animals were sufficient to protect against disease (Schiller and Lowy 1996).

These early animal studies of the L1 VLP approach were encouraging enough to prompt the NIH and several pharmaceutical companies to begin testing vaccine candidates in clinical studies (Schiller and Hidesheim 2000) that eventually led to key Phase III clinical trials with efficacy estimates ranging from 90 to 98% (Paavonen et al. 2007; The FUTURE II Study Group 2007). US licensure was granted by 2006 (quadrivalent Gardasil®) and 2009 (bivalent Cervarix®), with a 9-valent vaccine formulation (Gardasil 9®) licensed in 2014 (Centers for Disease Control and Prevention 2015d). Besides the number of HPV serotypes included in each formulation, the primary difference between the two vaccines is the inclusion of the monophosphoryl lipid A (MPL) adjuvant to the Cervarix vaccine in an effort to enhance antibody responses. In addition to providing impressive levels of protective efficacy, antibody responses to both vaccines appear to be long-lived. In one study, subjects vaccinated with three doses of either Cervarix or Gardasil were followed for up to 4 years (Einstein et al. 2014). While Cervarix induced higher serum and cervicovaginal neutralizing antibody titers, the rate of antibody decline was similar for both vaccines, with titers typically plateauing at 2–3 years and appearing stable thereafter (Einstein et al. 2014). Similar antibody kinetics have been observed by others (Giannini et al. 2006), with studies demonstrating both long-term antibody persistence (Safaeian et al. 2013) and efficacy following a single VLP vaccination (Kreimer et al. 2015). Although the precise mechanism of protection is unclear (Centers for Disease Control and Prevention 2015d), it is believed that serum neutralizing antibodies elicited by HPV vaccines can reach the genital tract through antibody exudation or transudation and thereby protect against infection at the mucosal surface (Schwarz and Leo 2008). Further evidence supporting this position comes from a recent study using a mouse cervicovaginal model which demonstrated that passive systemic transfer of immune serum alone could block HPV infection at the vaginal epithelium (Day et al. 2010). Unfortunately, despite proven efficacy, safety, and durability of immunity, only 57% of adolescent girls and 35% of adolescent boys in the USA had received a single HPV vaccine dose in 2013 (Stokley et al. 2014). This represents a significant missed opportunity, leaving a substantial number of young people at continued risk for infection and HPV-associated cancers.

3.3   Inactivated Measles and Respiratory Syncytial Virus Vaccines

The measles virus is one of the oldest recorded human pathogens, and despite the availability of an effective live, attenuated vaccine, it remains a significant threat for many parts of the world (World Health Organization 2014). The first live vaccine was licensed in the USA in 1963, and since that time the incidence of disease has declined dramatically (Centers for Disease Control and Prevention 2015e). In addition to the live vaccine, a formaldehyde-inactivated measles vaccine (FIMV) was also introduced in 1963 (Centers for Disease Control and Prevention 2015e). One trial involving over 5000 children indicated the FIMV could induce seroconversion, though immunity appeared to wane rapidly with titers declining to undetectable levels within 6 months following vaccination (Carter et al. 1962). Efficacy studies reinforced this result, with a reasonably high efficacy of 81% observed during the first 3 months post-vaccination, but dropping to only 63% by 11–13 months (Guinee et al. 1966). More concerning however, were multiple reports detailing an unusual and severe form of measles following wild-type exposure of children, all having previously received the inactivated vaccine (Fulginiti et al. 1967; Rauh and Schmidt 1965). This condition, termed atypical measles, presented a range of symptoms including high rates of muscle and abdominal pain as well as peripheral edema and pneumonia (Fulginiti et al. 1967; Rauh and Schmidt 1965). Based on this and other reports, the vaccine was removed from the market by 1967, just 4 years after its initial licensure (Centers for Disease Control and Prevention 2015e).

RSV was first described in the 1950s and was soon recognized as having a substantial impact on childhood respiratory health (Beem et al. 1960; Chanock et al. 1961). RSV remains a significant cause of acute lower respiratory infections among children <5 years of age, with a recent meta-analysis estimating an annual global burden of 33.8 million RSV-associated respiratory episodes and 66,000–199,000 deaths (Nair et al. 2010). Even in the USA, an estimated 2.1 million children require medical attention each year due to RSV, with ∼57,000 hospitalizations (Hall et al. 2009). Soon after the discovery of RSV, a formaldehyde-inactivated RSV (FIRSV) vaccine candidate was developed. However, multiple field efficacy trials demonstrated no protective immunity and instead provided evidence for increased risk for enhanced RSV disease in vaccine recipients (Chin et al. 1969; Fulginiti et al. 1969; Kapikian et al. 1969; Kim et al. 1969). For example, children vaccinated with either the FIRSV candidate or an inactivated parainfluenza vaccine as a control, demonstrated RSV attack rates that were similar between these groups (65 and 53%, respectively). However, the hospitalization rates among infected children in the FIRSV cohort reached 80% (compared to a rate of only 5% in the control group) and the vaccine was associated with two deaths (Kim et al. 1969).

The similarities between the failed measles virus and RSV-inactivated vaccines are evident (Polack 2007). Both pathogens are respiratory infections belonging to the *Paramyxoviridae* family, and while the underlying mechanism of immune enhancement of disease remains unclear, it is believed that inactivated vaccines against either virus fail to induce protective high-avidity antibody, with skewing toward a Th2 response and subsequent enhanced inflammation following wild-type infection (Acosta et al. 2016; Polack 2007). Importantly, it is also clear that neutralizing antibodies are key to controlling either disease. Passive immunotherapy for the prevention of measles was described as early as 1924 (Zingher 1924), with recent meta-analysis confirming its value (Young et al. 2014). Similarly, prophylactic administration of a RSV-neutralizing monoclonal antibody (palivizumab) has been shown to improve clinical outcomes and reduce hospitalizations in high-risk infants across multiple trials (Feltes et al. 2003; Pedraz et al. 2003; The IMpact-RSV Study Group 1998). These results suggest that protective immunity through vaccination is achievable for both pathogens, and this has been accomplished for measles through the use of a live, attenuated vaccine for >50 years. In contrast, while several new RSV vaccine candidates are under development, none have reached licensure (Jaberolansar et al. 2016). Nevertheless, recent positive results with a recombinant RSV F protein vaccine candidate in women (August et al. 2017; Glenn et al. 2016) provide some encouragement that the field is moving forward.

4 Successes and Failures of Subunit Vaccines

Subunit vaccines have been licensed for nine pathogens and encompass a range of approaches including purified toxoids, recombinant proteins, and either free or protein-conjugated polysaccharides (Table 1). In this section, we highlight the historic successes achieved with tetanus and diphtheria toxoids, both of which were developed in the 1920s (Roper et al. 2013; Tiwari and Wharton 2013). As a more recent example, we explore the development and implementation of successful vaccines against *Haemophilus influenza* type B (Hib) (Briere et al. 2014). These three examples have all benefited from having a defined antigen target with mechanisms of immunity established through passive immunization studies in animal models as well as human clinical practice. By comparison, the development of acellular pertussis (aP) vaccines faces ongoing challenges due to the waning immunity associated with this subunit vaccine approach (Plotkin 2014).

4.1   Tetanus and Diphtheria

Tetanus and diphtheria are both toxin-mediated diseases, against which effective vaccines have been widely available in the USA for >70 years (Roper et al. 2013; Tiwari and Wharton 2013). Tetanus is mediated by the neurotoxin tetanospasmin (i.e., tetanus toxin) expressed by the bacterium, *Clostridium tetani* (Roper et al. 2013), and diphtheria results from diphtheria toxin produced by *Corynebacterium diphtheriae* (Tiwari and Wharton 2013). Both vaccines consist of formaldehyde-inactivated toxoids with diphtheria toxoid developed in 1920 (Tiwari and Wharton 2013) and tetanus toxoid first described in 1924 (Roper et al. 2013). Efficacy of both toxoid vaccines has been demonstrated in large part by following epidemiology trends over time. Prior to broad introduction of tetanus vaccine in the 1940s, the USA recorded $\sim$500–600 cases of tetanus annually (Centers for Disease Control and Prevention 2015f). In the post-vaccination era, this rate has dropped to $\sim$30 cases annually and from 2001 to 2008 the CDC reported that in instances where vaccination status was reported, 41% of patients had never received a tetanus toxoid-containing vaccine (Centers for Disease Control and Prevention 2015f). Control of diphtheria (a communicable disease, in contrast to tetanus) has shown an even more dramatic impact. In the 1920s, there were an estimated 100,000–200,000 annual cases of diphtheria in the USA with up to 15,000 deaths (Centers for Disease Control and Prevention 2015b). These rates dropped rapidly after introduction of diphtheria vaccine, with no cases reported over an 11-year period from 2003 to 2013 (Adams et al. 2015). Immunity for both diseases is achieved through antibody-mediated toxin neutralization. For tetanus, immunity is considered 100% effective in preventing death if preexisting antibody levels are above a putative protective threshold of 0.01 IU/mL (Amanna et al. 2008). This protective threshold is bolstered both by direct human challenge studies performed in the 1940s (Wolters and Dehmel 1942) and analysis of patients admitted with clinical signs of tetanus, wherein only those patients with titers <0.01 IU/mL experience the most severe symptoms or die from the disease (Goulon et al. 1972). For diphtheria, a minimal protective serum titer has also been defined as 0.01 IU/mL (Bjorkholm et al. 1986; Tiwari and Wharton 2013).

The success of tetanus and diphtheria subunit vaccines may be viewed as a combination of several factors. Research into both toxins benefited from the development of relevant animal models and the demonstration that passive immunotherapy, directed against either toxin, was sufficient to achieve protective immunity (von Behring and Kitasato 1890). These results presented clear targets for vaccination with known mechanisms of protection (i.e., neutralizing antibodies). Furthermore, protection against both diseases can be achieved with relatively low levels of toxin-specific antibodies compared to the high levels of antibodies elicited through current vaccination schedules. For instance, following the completion of the primary tetanus and diphtheria vaccination series currently recommended in the USA (consisting of five doses through 4–6 years of age), average anti-tetanus and anti-diphtheria antibody titers reach >7 IU/mL (Black et al. 2006), several orders of magnitude above their protective thresholds of 0.01 IU/mL. In terms of longevity, we and others have shown that while antibody titers decline rapidly during the first few years following vaccination [typically 1–3 years following immunization (Amanna and Slifka 2010; Embree et al. 2015)], antitoxin antibodies then display longer-lived single-order kinetics, with tetanus-specific antibody half-lives ranging from 11 to 14 years (Amanna et al. 2007; Bonsignori et al. 2009; Hammarlund et al. 2016) and diphtheria-specific titers declining with an estimated half-life of 19–27 years (Amanna et al. 2007; Hammarlund et al. 2016). While these rates of decline are somewhat faster compared to other vaccine and viral antigens (Amanna et al. 2007), based on current antibody levels measured among US adults, at least 95% of the population will remain protected against both tetanus and diphtheria for 30 years or more without requiring further booster vaccination (Hammarlund et al. 2016). In light of these results, the current US recommendation for adult tetanus/diphtheria booster doses to be administered every 10 years (Kim et al. 2016) should be reconsidered.

4.2   Haemophilus Influenza Type B

Hib is a gram-negative bacterium that was once the leading cause of bacterial meningitis in US children (Briere et al. 2014). In the pre-vaccine era, annual US rates of invasive Hib in children <5 years old averaged ~20 cases per 100,000, but this has dropped dramatically to an average incidence of ~0.15 per 100,000 following licensure of effective vaccines in the late 1980s (Briere et al. 2014). Early passive immunotherapy studies in humans indicated antibodies could play an important role in controlling disease (Beck and Janney 1947), and these results were confirmed in later studies of at-risk children prophylactically treated with hyperimmune globulin (Santosham et al. 1990). Hib anti-capsular antibody was identified as the mechanism of protection as early as 1944 using a passive immunity mouse protection model (Alexander et al. 1944), and these studies were cited as a key motivation for the initial development and clinical testing of several capsular polysaccharide vaccine candidates (Anderson et al. 1972; Schneerson et al. 1971). This history demonstrates that, as with tetanus and diphtheria, vaccine development for Hib benefited from a clear antigen target for antibody-mediated protection and was an approach supported by passive immunity proof-of-principle studies carried out in both animals and humans.

The first Hib polysaccharide vaccine was licensed for use in the USA in 1985 (Briere et al. 2014). Pre-licensure studies had estimated vaccine efficacy of 90% (95% CI 58–99%) in 18–71-month-old children, but also demonstrated low antibody responses and no efficacy in children <18 months old (Peltola et al. 1984; Ward et al. 1988). Post-licensure evaluation through several case-control studies suggested more modest vaccine efficacy of only 50% in children $\geq$ 18 months of age (Ward et al. 1988). Immunogenicity was greatly enhanced with the development of polysaccharide–protein conjugate vaccines, with several formulations licensed for use in children as young as 2 months old by 1990 (Briere et al. 2014). Numerous controlled trials estimated the efficacy of the conjugate vaccines at $\sim$90% (Heath 1998), results which have been substantiated by the dramatic decrease of invasive Hib across multiple countries following widespread vaccination (Briere et al. 2014; Heath 1998). Current US recommendations suggest children should receive a 2–3 dose primary vaccination series (depending on the manufacturer) with a final booster dose typically given between 12 and 15 months of age (Briere et al. 2014). Following a similar vaccination schedule in UK infants, 98% of children were found to maintain Hib capsular polysaccharide antibody titers above the putative protective threshold [defined as 0.15 µg/mL (Anderson 1984)] for up to 2 years following the final booster vaccination (Borrow et al. 2010). A similar study in Spain indicated no drop in geometric mean antibody concentrations from 4.5 to 5.5 years following the final booster vaccination, with >97% of children maintaining antibody concentrations above the 0.15 µg/mL threshold (Tejedor et al. 2012). In total, these results indicate that the Hib conjugate vaccine can induce long-term antibody responses and effectively control this invasive and life-threatening disease.

4.3 Bordetella pertussis

Bordetella pertussis, the bacterial pathogen responsible for the disease commonly known as whooping cough, was at one time a leading cause of childhood morbidity and mortality in the USA (Morse 1913). In the decade just prior to vaccine licensure (1932–1941), annual reported attack rates averaged 157 per 100,000 persons, though the true rate has been estimated to be as high as 872 per 100,000 (Cherry 1999). Following the implementation of a whole-cell inactivated vaccine in the 1940s, disease burden dropped dramatically to a low of 0.5 cases per 100,000 persons by 1976 (Faulkner et al. 2015). While the positive public health impact of the whole-cell pertussis (wP) vaccine was striking, with estimates of vaccine efficacy ranging from 70 to 90% (Advisory Committee on Immunization Practices 1997), the vaccine elicited a relatively high rate of local and systemic adverse events (Cody et al. 1981). The most concerning adverse event was a potential link between vaccination and permanent brain damage due to encephalopathy

(Allen 2013; Shorvon and Berg 2008; Walton et al. 2015). However, several of the reported case studies supporting this hypothesis have since been rediagnosed as Dravet syndrome, a form of epilepsy associated with specific genetic mutations (Berkovic et al. 2006; Reyes et al. 2011), and reappraisal of earlier studies calls into question whether any link between wP vaccination and encephalopathy truly exists (Allen 2013; Shorvon and Berg 2008).

Nevertheless, given the public concern surrounding the safety of the wP vaccine, several replacement subunit acellular pertussis (aP) candidates were developed. Unlike tetanus, diphtheria, and Hib, one of the major limitations with the development of subunit vaccines for pertussis was the lack of a defined correlate of immunity against a specific *B. pertussis* component for vaccine-induced immunity. Early studies indicated a strong correlation between high serum agglutinin titers and protection from disease (Medical Research Council 1956; Sako 1947), but agglutinogens have been reported to include multiple targets such as pertactin (PRN), fimbriae-2 and 3 (FIM-2 and FIM-3), and lipooligosaccharide (LOS) (Cherry et al. 1998). Passive immunity studies performed in a mouse aerosol challenge model demonstrated that monoclonal antibodies against the pertussis toxin (PT), filamentous hemagglutinin (FHA), and FIM-2 could each protect mice against disease (Sato and Sato 1990), underscoring the complexity of the problem. Clinical results from early efficacy trials of aP vaccine candidates were likewise complex, suggesting that combinations of preexisting antibody titers to a range of antigens, including PT, PRN, and FIM-2, could be correlated with vaccine efficacy (Cherry et al. 1998; Storsaeter et al. 1998), but antibodies to FHA appeared to play no role in protection (Cherry et al. 1998) and no single measure seemed to assure protection.

Despite an uncertain understanding of the correlates of immunity associated with aP vaccines, several formulations were licensed for use in the USA by 1991 (Advisory Committee on Immunization Practices 1997) and contain a range of proteins including FHA, PRN, FIM-2, FIM-3, and inactivated pertussis toxoid. Current US recommendations for children include a five-dose schedule of DTaP (diphtheria, tetanus, acellular pertussis) through 4–6 years of age, with a Tdap booster (containing reduced content of both diphtheria and pertussis components) at 11–12 years of age (Faulkner et al. 2015). Initial estimates of vaccine efficacy appeared promising and ranged from 84 to 89%, and use of the aP vaccine was recommended for all vaccinations by 1997 (Advisory Committee on Immunization Practices 1997), with the wP vaccine phased out soon thereafter. However, in the decades following implementation of the full aP vaccine schedule, an alarming rise in pertussis cases was observed in several geographical regions including North America, Australia, and Europe (Plotkin 2014). Based on the close proximity in time between the resurgence of pertussis and the switch to the aP formulation, Australian researchers examined the relative risk for children developing pertussis based on their immunization history with either the wP or aP vaccines (Sheridan et al. 2012). Their study found that children who received a full aP vaccine schedule had a 3.29-fold increase (95% CI 2.44–4.46) in annual reporting rates for pertussis versus their wP-vaccinated counterparts. Furthermore, in mixed vaccination series (i.e., children receiving both wP and aP formulations) they found that the order of the vaccination series had a substantial impact on outcome, with aP primo-vaccinated children still demonstrating deficits in protection even when given the wP vaccine in later booster doses. This study has been supported by evidence from several US cohorts (Klein et al. 2013; Liko et al. 2013). In Oregon, analysis of children immunized during the transition from DTwP to DTaP demonstrated that priming with the acellular vaccine consistently increased the risk of contracting pertussis by ∼twofold across multiple vaccine booster scenarios (Liko et al. 2013). A case-control study performed in California indicated that teenagers who had received four doses of DTaP versus DTwP formulations during their initial vaccination series had a nearly sixfold increased risk of contracting PCR-positive pertussis (Klein et al. 2013).

While the underlying cause of poor protective efficacy is unclear, waning immunity may play a key role. A large-scale assessment of the five-dose DTaP schedule used in the USA estimated vaccine efficacy as a function of time (Misegades et al. 2012). This study found vaccine efficacy ranging from 92.3 to 98.1% during the first 3 years following the final booster dose, but this decreased over time to as low as 71.2% at time points >5 years post-vaccination. In 2006, the ACIP recommended adding a Tdap booster at 11–12 years of age in an attempt to counteract this waning immunity (Faulkner et al. 2015). However, boosting immunity in adolescents again appears to be limited. In California, researchers estimated vaccine efficacy of only 68.8% (95% CI 59.7–75.9) during the first year after Tdap vaccination of adolescents, dropping to as low as 8.9% (95% CI −30.6 to 36.4) by the fourth year (Klein et al. 2016). Studies in both Wisconsin (Koepke et al. 2014) and Washington (Acosta et al. 2015) substantiate these results, with vaccine efficacy ranging from 73.3 to 75% during the first year following vaccination, but rapidly waning to 11.9–34% within just a few years. The answer to this public health problem is unclear. A range of potential solutions has been proposed, and although a return to the old wP vaccine is considered unlikely (Plotkin 2014), new wP vaccines with high efficacy and improved safety profiles might still be a viable option. Increasing antigen doses may improve immunogenicity since differences in vaccine efficacy have been observed between different Tdap manufacturers, with higher antigen doses associated with increased vaccine efficacy (Koepke et al. 2014). However even with increased antigen dose, vaccine efficacy still wanes rapidly, pointing to what may be a fundamental limitation of the current acellular vaccine approach. Until researchers can find an improved approach to elicit durable immunity while retaining an acceptable safety profile, pertussis will represent a continuing problem for the foreseeable future.

5   Partially Successful Vaccines

In addition to examples of successful vaccines and clear vaccine failures, there are also a large number of partially successful vaccines. For the purposes of this discussion, we define a vaccine as "partially successful" if it elicits a protective immune response against a particular pathogen with efficacy rates that are near or substantially less than ~50%. Challenging pathogens, such as those with multiple serotypes, complex life cycles, or high antigenic variability, are the ones that are most likely to have vaccine counterparts that are either nonexistent or only partially successful. Some examples of challenging pathogens with partially successful vaccine candidates include (a) dengue virus (DENV) with its four interrelated serotypes, (b) VZV representing a latent herpesvirus that causes chicken pox (primary infection) or shingles (after reactivation/re-exposure), (c) *Plasmodium*, a parasite with a complex multistage life cycle that causes malaria, and (d) human immunodeficiency virus (HIV), a retrovirus with a complex life cycle as well as high antigenic variability.

5.1 Dengue Virus

DENV is a mosquito-borne pathogen that is ubiquitous throughout tropical regions of the world and endemic or epidemic in 154 countries (Furuya-Kanamori et al. 2016). DENV infection may result in a spectrum of disease ranging from acute, debilitating febrile illness (dengue fever, DF) to severe, life-threatening hemorrhagic disease (dengue hemorrhagic fever, DHF; dengue shock syndrome, DSS). Within DENV, there are four serotypes (DENV1, DENV2, DENV3, and DENV4), and each has been found to cause human disease and mortality. Current estimates indicate that nearly 4 billion people are at risk (Brady et al. 2012), with disease burden ranging from 58 to 96 million apparent DENV infections per year (Bhatt et al. 2013; Stanaway et al. 2016), resulting in approximately 500,000 cases of DHF/DSS and greater than 20,000 deaths (Murray et al. 2013). Infection with one DENV serotype is known to provide durable protective immunity against reinfection by that particular serotype (Webster et al. 2009). However, immunity to one DENV serotype may predispose an individual to more severe disease if infected by a second DENV serotype through a process believed to be associated with antibody-dependent enhancement (ADE) of infection of Fc-γ receptor-bearing cells (Burke et al. 1988; Guzman et al. 1987; Guzman et al. 2000; Halstead et al. 1970; Vaughn et al. 2000). Although the role of ADE in DHF/DSS has been questioned in recent years (Webster et al. 2009), this concept nonetheless remains an important consideration in terms of vaccine development. As such, it is widely accepted that an optimal DENV vaccine must provide protection against all four serotypes to provide broad antiviral immunity and to prevent the potential complications of ADE.

Dengvaxia® is a live, attenuated tetravalent vaccine that consists of four chimeric yellow fever virus strain 17D (YFV-17D) vectors expressing the DENV envelope and premembrane proteins of each DENV serotype, and has recently become the first licensed DENV vaccine (Sanofi Pasteur 2017). Despite the excitement surrounding this major milestone, concerns have emerged over efficacy and safety (Halstead and Russell 2016; Simmons 2015). An initial Phase IIb trial conducted in

Thailand estimated an overall vaccine efficacy of only 30.2% (95% CI −13.4 to 56.6), and the primary estimate of efficacy was not significant (Sabchareon et al. 2012). When segregated by serotype, there was some indication of efficacy for DENV1, DENV3, and DENV4, whereas with DENV2 (which constituted the majority of cases among vaccinated and controls) the vaccine demonstrated no protection. Larger Phase III trials in both Asia (Capeding et al. 2014) and Latin America (Villar et al. 2015) have now demonstrated modest protection against DENV, with vaccine efficacy ranging from 56.5 to 60.8%. However, as suggested in the Phase IIb trial, both studies demonstrated differences among serotypes, with DENV2 again showing the lowest protective immunity (35.0–42.3% vaccine efficacy). More concerning may be the limited protection observed in younger subjects and those who were seronegative at the time of vaccination. In the Asian trial, participant age ranged from 2 to 14 years old and decreasing efficacy were closely tied to decreasing age, with 2–5-year-old children demonstrating only 33.7% vaccine efficacy (95% CI 11.7–50.0) (Capeding et al. 2014). The Latin American trial only recruited children 9–16 years of age and found similar efficacy across this range (Villar et al. 2015). Across both Phase III trials, while vaccine efficacy in seropositive children ranged from 74.3 to 83.7%, it was sharply lower among previously seronegative children (vaccine efficacy = 35.5–43.2%) (Capeding et al. 2014; Villar et al. 2015). This suggests the age dependence observed in the Asian trial may be tied to the rates of seropositivity prior to vaccination within age groups (i.e., younger children have a lower likelihood of prior exposure to wild-type DENV). Seronegative children demonstrated neutralizing antibody titers that were typically ∼tenfold lower than their seropositive counterparts after vaccination [Table S8 in Villar et al. (2015)], and this limited immune response could play a role in the low observed vaccine efficacy. Unfortunately, this still leaves unexplained why seronegative children were unable to mount a robust immune response even after three immunizations with a live, attenuated vaccine spaced every 6 months. These results likely underlie the current recommendation to immunize only children ≥ 9 years of age (Sanofi Pasteur 2017). Unfortunately, this leaves a substantial gap in coverage, especially considering that younger children represent a vulnerable population that experience the highest rates of severe complications following DENV exposure (Guzman et al. 2002).

Ongoing long-term efficacy and safety analyses for Dengvaxia have been described (Arredondo-Garcia et al. 2018; Hadinegoro et al. 2015), with plans to monitor hospitalization rates for DENV illness in vaccinees and controls for up to 6 years. Interim analysis has indicated significant efficacy in children ≥ 9 years of age during years 1–2 post-vaccination, with a relative risk (RR) = 0.065 (95% CI: 0.01–0.22) among vaccinated children compared to controls (Arredondo-Garcia et al. 2018). However, vaccine efficacy waned rapidly, falling to nonsignificant levels by years 3–4 (RR = 0.752 [95% CI: 0.28–2.12]). Further, in children <9 years of age at the time of vaccination, no significant efficacy was observed at any time point following completion of the three-dose vaccination series. Additional analysis was performed comparing hospitalized cases of dengue in seropositive versus seronegative subjects at the time of vaccination across all age groups (Arredondo-Garcia et al. 2018). This analysis demonstrated a consistent benefit among seropositive subjects, with an overall RR = 0.292 (95% CI: 0.14–0.58). By comparison, pooling data from all seronegative subjects yielded a RR of 1.327 (95% CI: 0.56–3.48). Although the point estimate may suggest an increased disease incidence relative to controls, the wide confidence interval, which crosses the null value (i.e., RR = 1), indicates the result is not statistically significant. In addition to poor immunogenicity among seronegative children and the rapid loss of significant protection in any age group within 3–4 years, concerns have been raised over the potential for exacerbated disease and increased hospitalization of younger subjects (Hadinegoro et al. 2015). Whether or not this points to enhanced disease among children with suboptimal immune responses remains uncertain, although one study noted that the frequency of severe forms of DENV (such as DSS) does not appear to be increased in this age group (Simmons 2015). Nevertheless, given the limited efficacy of the vaccine, substantial age restrictions, and continuing safety concerns, DENV will remain a global threat with limited solutions.

5.2 Varicella Zoster Virus (Live, Attenuated)

The live, attenuated VZV vaccine, originally developed for the prevention of varicella (i.e., chicken pox), was initially believed to require only one dose to achieve full protective immunity (Izurieta et al. 1997; Lopez et al. 2006; Vazquez et al. 2001) and was licensed using this schedule in 1995. However, additional studies demonstrated that protective immunity waned over time and a two-dose schedule for this live vaccine was incorporated into the ACIP recommendations by 2007 (Marin et al. 2007). Similarly, it was originally assumed that a single dose of live varicella vaccine would be successful for protection of older individuals (≥ 60 years of age) against herpes zoster (i.e., shingles). However, in a pivotal randomized double-blinded clinical trial involving 38,546 adults monitored for a median of 3.12 years, the Oka/Merck VZV vaccine (Zostavax®) provided 51.3% efficacy against clinical diagnosis of herpes zoster (Oxman et al. 2005). Vaccine efficacy was higher among younger subjects (63.9%) but reached only 37.6% efficacy in subjects 70 years of age or older—indicating that the vaccine performed poorly in the aged population at most risk for developing severe herpes zoster and post-herpetic neuralgia. In 2006, the ACIP recommended a single dose of vaccine against herpes zoster (Harpaz et al. 2008), but in 2013 the ACIP declined to recommend the vaccine for younger adults 50–59 years of age due to the limited amount of information regarding the durability of protection after vaccination (Tseng et al. 2016). Indeed, several studies found evidence of rapidly declining protection (Morrison et al. 2015; Schmader et al. 2012; Tseng et al. 2016) with the most compelling evidence presented in a study comparing 176,078 vaccinated subjects to 528,234 unvaccinated subjects who were followed for up to 8 years after vaccination (Tseng et al. 2016). Although vaccine efficacy approached 69% during the first year after vaccination, protective immunity declined rapidly thereafter with just 4% vaccine efficacy by the eighth year after immunization. The decline in protection against disease was similar among vaccine recipients aged 60–69 years and those aged ≥ 70 years, indicating that advanced age itself was not responsible for the decline in protective immunity.

Although no proven correlate of protective immunity exists for herpes zoster, a long-held belief has been that loss of antiviral T cell-mediated immunity is the determining risk factor for disease onset (Kost and Straus 1996; Miller 1980; Wilson et al. 1992). Along these lines, studies demonstrated that VZV vaccination resulted in a substantial increase in T cell-mediated immunity (Levin et al. 2003; Oxman 1995) and the dose of vaccine for prevention of herpes zoster (approximately 14-fold higher than the dose administered during primary vaccination to prevent varicella) was chosen on the basis of its ability to elicit a significant increase in antiviral T cell immunity among older adults (Oxman et al. 2005). Bearing in mind that the first-line approach of using a live, attenuated strain of VZV for vaccination would be the most likely candidate for eliciting strong antiviral T cell responses, one would have found little support for the use of a non-replicating subunit vaccine since it would presumably be far less likely to elicit the T cell responses that develop following infection with a live, attenuated version of the same virus. Nevertheless, conventional wisdom was displaced by the success of a non-traditional approach to vaccination against herpes zoster with a subunit vaccine consisting of a single viral protein antigen, recombinant VZV glycoprotein E (gE) (Shingrix®) (Lal et al. 2015). In this study involving 15,411 subjects (7698 vaccinated and 7713 placebo controls) who were followed for a mean of 3.2 years, vaccine efficacy was 97.2%. There was also no age-associated reduction in vaccine efficacy among subjects ≥ 70 years of age. This indicated that the immunosenescence and age-associated reduction in vaccine efficacy observed after live VZV vaccination could be overcome with a simple two-dose regimen of a non-replicating subunit vaccine. Further analysis of the subunit vaccine-induced VZV gE-specific $CD4^+$ T cell responses and gE-specific antibody responses indicates that immunity initially declines for the first 1–2 years after vaccination but then reaches a plateau that is maintained for at least 6 years. In these studies, $CD4^+$ T cell memory was sustained at nearly fourfold higher than pre-vaccination levels and antiviral antibody responses remain approximately sevenfold higher than pre-vaccination levels (Chlibek et al. 2016). Together, these results indicate that this simple subunit vaccine has the potential to provide long-lasting immunity against herpes zoster. This vaccine was approved by the FDA in 2017, and given its superior and longer-lasting efficacy, the ACIP recommended Shingrix as the preferred herpes zoster vaccine (Dooling et al. 2018).

5.3 *Plasmodium*

Plasmodium parasites are the causative agent of malaria, and the complex life cycle of this parasite has been problematic for the development of an effective vaccine.

The human host is infected by sporozoites transmitted by infected mosquitos. The sporozoites travel through the bloodstream and infect hepatocytes where they reproduce as merozoites. The merozoites infect and multiply in new red blood cells that lyse to release more merozoites and initiate the next round of infection. Some merozoites develop into the precursors of male and female gametes, which can infect a feeding mosquito and mature, fuse to form fertilized ookinetes, and develop into sporozoites that travel to the salivary glands where they can be transmitted to the next human host during blood feeding. The invading sporozoites that are transmitted by infected mosquitos represent the best vaccine target since blocking infection at the earliest stage of transmission is likely to provide the highest clinical benefit. The most advanced malaria vaccine candidate is RTS,S, a subunit vaccine based on a *Plasmodium* circumsporozoite protein (CSP) that is genetically fused to hepatitis B virus surface antigen. The vaccine has been in development for more than 30 years (Kaslow and Biernaux 2015) and has advanced through Phase III clinical trials. Protection against clinical malaria in children (ages 5–17 months) after three doses of vaccine is modest; vaccine efficacy reached 45.1% during the first 0–20 months after vaccination but waned rapidly, with only 16.1% vaccine efficacy observed during months 21–32 in young children. The levels of protection are even lower among infants (ages 6–12 weeks) with only 27.0% protective efficacy at 0–20 months with immunity falling to near negligible levels (7.6%) from 21 to 32 months after vaccination (RTS-S Clinical Trials Partnership 2015). The partial success among children and the weak protective immunity elicited among infants indicates that this subunit vaccine is unable to reliably protect the most vulnerable populations who are at the highest risk for severe cases of malaria and this is compounded by the lack of durable immunity even after a three-dose regimen.

An alternative approach to malaria vaccination that has gained recent attention utilizes intact gamma-irradiated sporozoites purified from infected mosquitos (Hoffman et al. 2015; Richie et al. 2015). The main advantage of this approach is that vaccination is performed using the whole organism in which many potentially immunogenic proteins are presented in their native conformation. In a human challenge model of *Plasmodium* infection, the highest administered dose provided 100% protection from parasitemia for six immunized subjects (Seder et al. 2013). The main disadvantages to this approach are that direct intravenous injection of the vaccine is required [early studies involving subcutaneous or intradermal vaccination failed (Epstein et al. 2011)] and there are technical limitations to the production, purification, and long-term stability of sporozoites prepared from *Plasmodium*-infected mosquitos. Another potential concern is that the vaccine may be on the knife's edge in terms of protective efficacy since a four-dose vaccination schedule elicited promastigote-specific antibodies that were roughly half of that obtained from a five-dose schedule and similarly protection from clinical disease declined from 100% (6/6 subjects) to 60% (6/9 subjects). Since many vaccine-mediated immune responses decline by two- to tenfold from their peak levels before reaching a plateau stage of more stable long-term maintenance, it is possible that this vaccine approach may lose at least some protective efficacy over a relatively short period of time. Nevertheless, the results of these studies provide a foundation on which future vaccines may be developed that elicit appropriate immune responses that block the earliest stages of *Plasmodium* infection and further studies are currently underway (Hoffman et al. 2015).

5.4   *Human Immunodeficiency Virus*

HIV, the causative agent of acquired immune deficiency syndrome (AIDS), is a notoriously difficult vaccine target, and of the six field efficacy trials that have been performed to date, all HIV vaccines have failed except one, the RV144 vaccine (Kim et al. 2015). The RV144 trial involved 16,402 subjects and a prime–boost strategy utilizing a recombinant canarypox vector expressing HIV Gag, pro, and envelope administered at 0, 1, 3, and 6 months in addition to alum-adjuvanted monomeric envelope proteins co-administered at the 3- and 6-month time points. Vaccine efficacy reached 60.5% at 1 year after vaccination but declined to 31.2% efficacy at 3.5 years after vaccination (Kim et al. 2015). Despite these suboptimal results, any significant protection against HIV is encouraging and current clinical trials involve redesigned gp120 envelope constructs, comparison of different adjuvants (MF59 and AS01B), and a revised vaccination schedule that includes a booster dose administered at 12 months (Gray et al. 2016). Moreover, development of intact HIV envelope trimers provides a new approach to HIV vaccine development and further studies are planned to test multiple trimers in either simultaneous or sequential combinations in order to optimally induce high titer, broadly neutralizing antibodies (de Taeye et al. 2016). Regardless of the approach that is taken, development of a vaccination strategy that induces long-term immunity above a protective threshold [preferably with broadly neutralizing antibodies (Burton and Mascola 2015)] is a high priority and likely to be key to the success of any future HIV vaccine.

6   Future Vaccines for Challenging Pathogens

Despite the ultimate goal of performing rational vaccine design in which a deeper understanding of pathogen structure, immunogenicity, and pathogenesis is utilized in the construction of a new vaccine, many of our current successful vaccine strategies have been developed empirically. In some cases such as measles, mumps, and rubella, an attenuated version of the target virus provides safe and effective vaccine-mediated protection. In other cases, successful vaccine development was counterintuitive to the approach that one might have predicted based on the characteristics of the pathogen itself. HPV, for example, represents a mucosal pathogen that evades the immune system in order to sustain a chronic essentially lifelong infection within its human host. One might have predicted that a mucosal route of vaccination would be preferred or that CD8$^+$ T cells would be the most important cell type for protecting against intracellular pathogens and that antiviral IgA responses might be much more important than IgG responses in terms of protecting mucosal sites from natural infection. In contrast to these a priori expectations, development of non-replicating HPV VLP vaccines that are administered by intramuscular injection induce strong IgG responses and relatively minor IgA or T cell responses but still provide sustained protective immunity against HPV for many years after vaccination. This result would not have been predicted based on conventional wisdom and knowledge of the natural host–pathogen interactions and provides an interesting counterpoint showing that sometimes it is important to take empirical approaches to the development of vaccines to the most complex pathogens.

Acknowledgements This work was funded in part with federal funds from the National Institute of Allergy and Infectious Diseases R44 AI079898 (to MKS and IJA), R01 AI098723 (to MKS), and Oregon National Primate Research Center grant, P51OD011092 (to MKS). OHSU and Dr. Slifka have a financial interest in Najít Technologies, Inc., a company that may have a commercial interest in the results of this research and technology. This potential individual and institutional conflict of interest have been reviewed and managed by OHSU. Dr. Amanna is an employee of Najít Technologies, Inc. No writing assistance was utilized in the production of this manuscript.

References

Acosta AM et al (2015) Tdap vaccine effectiveness in adolescents during the 2012 Washington State pertussis epidemic. Pediatrics 135:981–989. https://doi.org/10.1542/peds.2014-3358

Acosta PL, Caballero MT, Polack FP (2016) Brief history and characterization of enhanced respiratory syncytial virus disease. Clin Vaccine Immunol 23:189–195. https://doi.org/10.1128/CVI.00609-15

Adams D et al (2015) Summary of notifiable infectious diseases and conditions—United States, 2013 MMWR Morb Mortal Wkly Rep 62:1–122. https://doi.org/10.15585/mmwr.mm6253a1

Advisory Committee on Immunization Practices (1997) Pertussis vaccination: use of acellular pertussis vaccines among infants and young children. MMWR Recomm Rep 46:1–32

Advisory Committee on Immunization Practices (2000) Poliomyelitis prevention in the United States. Updated recommendations of the Advisory Committee on Immunization Practices (ACIP). MMWR Recomm Rep 49:1–22

Alexander HE, Heidelberger M, Leidy G (1944) The protective or curative element in type B H. influenzae rabbit serum. Yale J Biol Med 16:425–434

Alexander LN et al (2004) Vaccine policy changes and epidemiology of poliomyelitis in the United States. JAMA 292:1696–1701

Allen A (2013) Public health. The pertussis paradox. Science 341:454–455. https://doi.org/10.1126/science.341.6145.454

Amanna IJ, Carlson NE, Slifka MK (2007) Duration of humoral immunity to common viral and vaccine antigens. N Engl J Med 357:1903–1915

Amanna IJ, Messaoudi I, Slifka MK (2008) Protective immunity following vaccination: how is it defined? Hum Vaccine 4:316–319 (5751 [pii])

Amanna IJ, Slifka MK (2010) Mechanisms that determine plasma cell lifespan and the duration of humoral immunity. Immunol Rev 236:125–138. https://doi.org/10.1111/j.1600-065X.2010.00912.x (IMR912 [pii])

Anderson P (1984) The protective level of serum antibodies to the capsular polysaccharide of Haemophilus influenzae type b. J Infect Dis 149:1034–1035

Anderson P, Johnston RB Jr, Smith DH (1972) Human serum activities against Hemophilus influenzae, type b. J Clin Invest 51:31–38. https://doi.org/10.1172/JCI106793

Arredondo-Garcia JL et al (2018) Four-year safety follow-up of the tetravalent dengue vaccine efficacy randomised controlled trials in Asia and Latin America. Clin Microbiol Infect. https://doi.org/10.1016/j.cmi.2018.01.018

August A et al (2017) A phase 2 randomized, observer-blind, placebo-controlled, dose-ranging trial of aluminum-adjuvanted respiratory syncytial virus F particle vaccine formulations in healthy women of childbearing age. Vaccine 35:3749–3759. https://doi.org/10.1016/j.vaccine.2017.05.045

Bacungan VJ, de Guzman C, Hermonio E (2018) TIMELINE: The Dengvaxia controversy. CNN Philippines. http://cnnphilippines.com/news/2017/12/09/The-Dengvaxia-controversy.html. Accessed 12 Feb 2018

Beck KH, Janney FR (1947) Alexander's rabbit serum in the treatment of influenzal meningitis; an evaluation of its use in conjunction with sulfonamide compounds. JAMA 73:317–325

Beem M, Wright FH, Hamre D, Egerer R, Oehme M (1960) Association of the chimpanzee coryza agent with acute respiratory disease in children. N Engl J Med 263:523–530. https://doi.org/10.1056/NEJM196009152631101

Berkovic SF et al (2006) De-novo mutations of the sodium channel gene SCN1A in alleged vaccine encephalopathy: a retrospective study. Lancet Neurol 5:488–492. https://doi.org/10.1016/S1474-4422(06)70446-X

Bhatt S et al (2013) The global distribution and burden of dengue. Nature 496:504–507. https://doi.org/10.1038/nature12060

Bjorkholm B, Bottiger M, Christenson B, Hagberg L (1986) Antitoxin antibody levels and the outcome of illness during an outbreak of diphtheria among alcoholics. Scand J Infect Dis 18:235–239

Black S, Friedland LR, Schuind A, Howe B (2006) Immunogenicity and safety of a combined DTaP-IPV vaccine compared with separate DTaP and IPV vaccines when administered as pre-school booster doses with a second dose of MMR vaccine to healthy children aged 4–6 years. Vaccine 24:6163–6171 https://doi.org/10.1016/j.vaccine.2006.04.001

Bonsignori M et al (2009) HIV-1 envelope induces memory B cell responses that correlate with plasma antibody levels after envelope gp120 protein vaccination or HIV-1 infection. J Immunol 183:2708–2717. https://doi.org/10.4049/jimmunol.0901068

Borrow R et al (2010) Kinetics of antibody persistence following administration of a combination meningococcal serogroup C and haemophilus influenzae type b conjugate vaccine in healthy infants in the United Kingdom primed with a monovalent meningococcal serogroup C vaccine. Clin Vaccine Immunol 17:154–159. https://doi.org/10.1128/CVI.00384-09

Brady OJ et al (2012) Refining the global spatial limits of dengue virus transmission by evidence-based consensus. PLoS Negl Trop Dis 6:e1760. https://doi.org/10.1371/journal.pntd.0001760

Briere EC, Rubin L, Moro PL, Cohn A, Clark T, Messonnier N (2014) Prevention and control of haemophilus influenzae type b disease: recommendations of the advisory committee on immunization practices (ACIP). MMWR Recomm Rep 63:1–14

Burke DS, Nisalak A, Johnson DE, Scott RM (1988) A prospective study of dengue infections in Bangkok. Am J Trop Med Hyg 38:172–180

Burton DR, Mascola JR (2015) Antibody responses to envelope glycoproteins in HIV-1 infection. Nat Immunol 16:571–576. https://doi.org/10.1038/ni.3158ni.3158 [pii]

Capeding MR et al (2014) Clinical efficacy and safety of a novel tetravalent dengue vaccine in healthy children in Asia: a phase 3, randomised, observer-masked, placebo-controlled trial. Lancet 384:1358–1365. https://doi.org/10.1016/S0140-6736(14)61060-6

Cardemil CV et al (2017) Effectiveness of a third dose of MMR vaccine for mumps outbreak control. N Engl J Med 377:947–956. https://doi.org/10.1056/NEJMoa1703309

Carter CH et al (1962) Serologic response of children to in-activated measles vaccine. JAMA 179:848–853

Centers for Disease Control and Prevention (1996) Prevention of hepatitis A through active or passive immunization: recommendations of the Advisory Committee on Immunization Practices (ACIP). MMWR Recomm Rep 45:1–30

Centers for Disease Control and Prevention (2015a) Cervical cancer statistics. Centers for disease control and prevention. http://www.cdc.gov/cancer/cervical/statistics/. Accessed 18 Apr 2016

Centers for Disease Control and Prevention (2015b) Diphtheria. In: Hamborsky J, Kroger A, Wolfe S (eds) Epidemiology and prevention of vaccine-preventable diseases, 13th edn. Public Health Foundation, Washington, D.C., pp 107–118

Centers for Disease Control and Prevention (2015c) Hepatitis A. In: Hamborsky J, Kroger A, Wolfe S (eds) Epidemiology and prevention of vaccine-preventable diseases, 13th edn. Public Health Foundation, Washington, D.C., pp 135–148

Centers for Disease Control and Prevention (2015d) Human papillomavirus. In: Hamborsky J, Kroger A, Wolfe S (eds) Epidemiology and prevention of vaccine-preventable diseases, 13th edn. Public Health Foundation, Washington, D.C., pp 175–186

Centers for Disease Control and Prevention (2015e) Measles. In: Hamborsky J, Kroger A, Wolfe S (eds) Epidemiology and prevention of vaccine-preventable diseases, 13th edn. Public Health Foundation, Washington, D.C., pp 209–230

Centers for Disease Control and Prevention (2015f) Tetanus. In: Hamborsky J, Kroger A, Wolfe S (eds) Epidemiology and prevention of vaccine-preventable diseases, 13th edn. Public Health Foundation, Washington, D.C., pp 341–352

Chanock RM, Kim HW, Vargosko AJ, Deleva A, Johnson KM, Cumming C, Parrott RH (1961) Respiratory syncytial virus. I. Virus recovery and other observations during 1960 outbreak of bronchiolitis, pneumonia, and minor respiratory diseases in children. JAMA 176:647–653

Cherry JD (1999) Pertussis in the preantibiotic and prevaccine era, with emphasis on adult pertussis. Clin Infect Dis 28(Suppl 2):S107–S111. https://doi.org/10.1086/515057

Cherry JD, Gornbein J, Heininger U, Stehr K (1998) A search for serologic correlates of immunity to *Bordetella pertussis* cough illnesses. Vaccine 16:1901–1906

Chin J, Magoffin RL, Shearer LA, Schieble JH, Lennette EH (1969) Field evaluation of a respiratory syncytial virus vaccine and a trivalent parainfluenza virus vaccine in a pediatric population. Am J Epidemiol 89:449–463

Chlibek R et al (2016) Long-term immunogenicity and safety of an investigational herpes zoster subunit vaccine in older adults. Vaccine 34:863–868. https://doi.org/10.1016/j.vaccine.2015.09.073s0264-410x(15)01349-3 [pii]

Cody CL, Baraff LJ, Cherry JD, Marcy SM, Manclark CR (1981) Nature and rates of adverse reactions associated with DTP and DT immunizations in infants and children. Pediatrics 68:650–660

Day PM, Kines RC, Thompson CD, Jagu S, Roden RB, Lowy DR, Schiller JT (2010) In vivo mechanisms of vaccine-induced protection against HPV infection. Cell Host Microbe 8:260–270. https://doi.org/10.1016/j.chom.2010.08.003

de Taeye SW, Moore JP, Sanders RW (2016) HIV-1 Envelope trimer design and immunization strategies to induce broadly neutralizing antibodies. Trends Immunol 37:221–232. https://doi.org/10.1016/j.it.2016.01.007s1471-4906(16)00008-9 [pii]

Dooling KL, Guo A, Patel M, Lee GM, Moore K, Belongia EA, Harpaz R (2018) Recommendations of the Advisory Committee on Immunization Practices for Use of Herpes Zoster Vaccines. MMWR Morb Mortal Wkly Rep 67:103–108. https://doi.org/10.15585/mmwr.mm6703a5

Einstein MH et al (2014) Comparative humoral and cellular immunogenicity and safety of human papillomavirus (HPV)-16/18 AS04-adjuvanted vaccine and HPV-6/11/16/18 vaccine in healthy women aged 18–45 years: follow-up through month 48 in a phase III randomized study. Hum Vaccin Immunother 10:3455–3465. https://doi.org/10.4161/hv.36117

Embree J, Law B, Voloshen T, Tomovici A (2015) Immunogenicity, safety, and antibody persistence at 3, 5, and 10 years postvaccination in adolescents randomized to booster immunization with a combined tetanus, diphtheria, 5-component acellular pertussis, and inactivated poliomyelitis vaccine administered with a hepatitis B virus vaccine concurrently or 1 month apart. Clin Vaccine Immunol 22:282–290. https://doi.org/10.1128/CVI.00682-14

Epstein JE et al (2011) Live attenuated malaria vaccine designed to protect through hepatic CD8 (+) T cell immunity. Science 334:475–480. https://doi.org/10.1126/science.1211548science. 1211548 [pii]

Faulkner A, Skoff T, Martin S, Cassiday P, Tondella ML, Liang J (2015) Chapter 10: Pertussis. In: Manual for the surveillance of vaccine-preventable diseases. Centers for Disease Control and Prevention, Atlanta, GA

Feltes TF et al (2003) Palivizumab prophylaxis reduces hospitalization due to respiratory syncytial virus in young children with hemodynamically significant congenital heart disease. J Pediatr 143:532–540

Fulginiti VA, Eller JJ, Downie AW, Kempe CH (1967) Altered reactivity to measles virus. Atypical measles in children previously immunized with inactivated measles virus vaccines. JAMA 202:1075–1080

Fulginiti VA, Eller JJ, Sieber OF, Joyner JW, Minamitani M, Meiklejohn G (1969) Respiratory virus immunization. I. A field trial of two inactivated respiratory virus vaccines; an aqueous trivalent parainfluenza virus vaccine and an alum-precipitated respiratory syncytial virus vaccine. Am J Epidemiol 89:435–448

Furuya-Kanamori L et al (2016) Co-distribution and co-infection of chikungunya and dengue viruses. BMC Infect Dis 16:84. https://doi.org/10.1186/s12879-016-1417-2

Garon J, Patel M (2017) The polio endgame: rationale behind the change in immunisation. Arch Dis Child 102:362–365. https://doi.org/10.1136/archdischild-2016-311171

Gellis SS, Stokes J Jr, Brother GM, Hall WM, Gilmore HR, Beyer E, Morrissey RA (1945) The use of human immune serum globulin (gamma globulin) in infectious (epidemic) hepatitis in the Mediterranean theater of operations I. Studies on prophylaxis in two epidemics of infectious hepatitis. JAMA 128:1062–1063

Giannini SL et al (2006) Enhanced humoral and memory B cellular immunity using HPV16/18 L1 VLP vaccine formulated with the MPL/aluminium salt combination (AS04) compared to aluminium salt only. Vaccine 24:5937–5949. https://doi.org/10.1016/j.vaccine.2006.06.005 (S0264-410X(06)00709-2 [pii])

Glenn GM et al (2016) A randomized, blinded, controlled, dose-ranging study of a respiratory syncytial virus recombinant fusion (F) nanoparticle vaccine in healthy women of childbearing age. J Infect Dis 213:411–422. https://doi.org/10.1093/infdis/jiv406

Goulon M, Girard O, Grosbuis S, Desormeau JP, Capponi MF (1972) Antitetanus antibodies. Assay before anatoxinotherapy in 64 tetanus patients. Nouv Presse Med 1:3049–3050

Gray GE, Laher F, Lazarus E, Ensoli B, Corey L (2016) Approaches to preventative and therapeutic HIV vaccines. Curr Opin Virol 17:104–109. https://doi.org/10.1016/j.coviro.2016. 02.010 (S1879-6257(16)30014-1 [pii])

Grohskopf LA, Sokolow LZ, Broder KR, Walter EB, Bresee JS, Fry AM, Jernigan DB (2017) Prevention and control of seasonal influenza with vaccines: recommendations of the advisory committee on immunization practices—United States, 2017–18 Influenza Season. MMWR Recomm Rep 66:1–20. https://doi.org/10.15585/mmwr.rr6602a1

Guinee VF et al (1966) Cooperative measles vaccine field trial. I. Clinical efficacy. Pediatrics 37:649–665

Guzman MG, Kouri G, Bravo J, Valdes L, Vazquez S, Halstead SB (2002) Effect of age on outcome of secondary dengue 2 infections. Int J Infect Dis 6:118–124

Guzman MG et al (1987) Clinical and serologic study of Cuban children with dengue hemorrhagic fever/dengue shock syndrome (DHF/DSS). Bull Pan Am Health Organ 21:270–279

Guzman MG et al (2000) Epidemiologic studies on dengue in Santiago de Cuba, 1997. Am J Epidemiol 152:793–799 (discussion 804)

Hadinegoro SR et al (2015) Efficacy and long-term safety of a dengue vaccine in regions of endemic disease. N Engl J Med 373:1195–1206. https://doi.org/10.1056/NEJMoa1506223

Hall CB et al (2009) The burden of respiratory syncytial virus infection in young children. N Engl J Med 360:588–598. https://doi.org/10.1056/NEJMoa0804877

Halstead SB, Nimmannitya S, Cohen SN (1970) Observations related to pathogenesis of dengue hemorrhagic fever. IV. Relation of disease severity to antibody response and virus recovered. Yale J Biol Med 42:311–328

Halstead SB, Russell PK (2016) Protective and immunological behavior of chimeric yellow fever dengue vaccine. Vaccine 34:1643–1647. https://doi.org/10.1016/j.vaccine.2016.02.004

Hammarlund E et al (2016) Durability of vaccine-induced immunity against tetanus and diphtheria toxins: a cross-sectional analysis. Clin Infect Dis 62:1111–1118. https://doi.org/10.1093/cid/ciw066

Hao L, Toyokawa S, Kobayashi Y (2008) Poisson-model analysis of the risk of vaccine-associated paralytic poliomyelitis in Japan between 1971 and 2000. Jpn J Infect Dis 61:100–103

Harpaz R, Ortega-Sanchez IR, Seward JF (2008) Prevention of herpes zoster: recommendations of the Advisory Committee on Immunization Practices (ACIP). MMWR Recomm Rep 57:1–30 (quiz CE32-34 doi:rr5705a1 [pii])

Havens WP Jr, Paul JR (1945) Prevention of infectious hepatitis with gamma globulin. JAMA 129:270–272

Heath PT (1998) Haemophilus influenzae type b conjugate vaccines: a review of efficacy data. Pediatr Infect Dis J 17:S117–S122

Heinz FX, Stiasny K, Holzmann H, Grgic-Vitek M, Kriz B, Essl A, Kundi M (2013) Vaccination and tick-borne encephalitis, central Europe. Emerg Infect Dis 19:69–76. https://doi.org/10.3201/eid1901.120458

Hoffman SL, Vekemans J, Richie TL, Duffy PE (2015) The march toward malaria vaccines. Vaccine 33(Suppl 4):D13–D23 https://doi.org/10.1016/j.vaccine.2015.07.091s0264-410x(15)01070-1 [pii]

Hsia DY, Lonsway M Jr, Gellis SS (1954) Gamma globulin in the prevention of infectious hepatitis; studies on the use of small doses in family outbreaks. N Engl J Med 250:417–419. https://doi.org/10.1056/NEJM195403112501004

Innis BL et al (1994) Protection against hepatitis A by an inactivated vaccine. JAMA 271:1328–1334

Izurieta HS, Strebel PM, Blake PA (1997) Postlicensure effectiveness of varicella vaccine during an outbreak in a child care center. JAMA 278:1495–1499

Jaberolansar N, Toth I, Young PR, Skwarczynski M (2016) Recent advances in the development of subunit-based RSV vaccines. Expert Rev Vaccines 15:53–68. https://doi.org/10.1586/14760584.2016.1105134

Jenner E (1798) An inquiry into the causes and effects of the variolae vaccinae. Sampson Low, London

Jenner E (1799) Further observations on the variolae vaccinae. Sampson Low, London

Jenner E (1800) A continuation of facts and observations relative to the variolae vaccinae, or cowpox. Sampson Low, London

Kapikian AZ, Mitchell RH, Chanock RM, Shvedoff RA, Stewart CE (1969) An epidemiologic study of altered clinical reactivity to respiratory syncytial (RS) virus infection in children previously vaccinated with an inactivated RS virus vaccine. Am J Epidemiol 89:405–421

Kaslow DC, Biernaux S (2015) RTS,S: Toward a first landmark on the malaria vaccine technology roadmap. Vaccine 33:7425–7432. https://doi.org/10.1016/j.vaccine.2015.09.061s0264-410x(15)01337-7 [pii]

Kim DK, Bridges CB, Harriman KH (2016) Advisory Committee on Immunization Practices AAIWG (2016) Advisory Committee on Immunization Practices recommended immunization schedule for adults aged 19 years or older—United States. MMWR Morb Mortal Wkly Rep 65:88–90. https://doi.org/10.15585/mmwr.mm6504a5

Kim HW, Canchola JG, Brandt CD, Pyles G, Chanock RM, Jensen K, Parrott RH (1969) Respiratory syncytial virus disease in infants despite prior administration of antigenic inactivated vaccine. Am J Epidemiol 89:422–434

Kim JH, Excler JL, Michael NL (2015) Lessons from the RV144 Thai phase III HIV-1 vaccine trial and the search for correlates of protection. Annu Rev Med 66:423–437. https://doi.org/10.1146/annurev-med-052912-123749

Kirnbauer R, Booy F, Cheng N, Lowy DR, Schiller JT (1992) Papillomavirus L1 major capsid protein self-assembles into virus-like particles that are highly immunogenic. Proc Natl Acad Sci USA 89:12180–12184

Klein NP, Bartlett J, Fireman B, Baxter R (2016) Waning Tdap effectiveness in adolescents. Pediatrics 137:1–9. https://doi.org/10.1542/peds.2015-3326

Klein NP, Bartlett J, Fireman B, Rowhani-Rahbar A, Baxter R (2013) Comparative effectiveness of acellular versus whole-cell pertussis vaccines in teenagers. Pediatrics 131:e1716–e1722. https://doi.org/10.1542/peds.2012-3836

Koepke R et al (2014) Estimating the effectiveness of tetanus-diphtheria-acellular pertussis vaccine (Tdap) for preventing pertussis: evidence of rapidly waning immunity and difference in effectiveness by Tdap brand. J Infect Dis 210:942–953. https://doi.org/10.1093/infdis/jiu322

Kost RG, Straus SE (1996) Postherpetic neuralgia–pathogenesis, treatment, and prevention. N Engl J Med 335:32–42. https://doi.org/10.1056/NEJM199607043350107

Kramarz P et al (2001) Population-based study of rotavirus vaccination and intussusception. Pediatr Infect Dis J 20:410–416

Kreimer AR et al (2015) Efficacy of fewer than three doses of an HPV-16/18 AS04-adjuvanted vaccine: combined analysis of data from the Costa Rica Vaccine and PATRICIA Trials. Lancet Oncol 16:775–786. https://doi.org/10.1016/S1470-2045(15)00047-9

Lal H et al (2015) Efficacy of an adjuvanted herpes zoster subunit vaccine in older adults. N Engl J Med 372:2087–2096. https://doi.org/10.1056/NEJMoa1501184

Lee LA et al (2013) The estimated mortality impact of vaccinations forecast to be administered during 2011–2020 in 73 countries supported by the GAVI alliance. Vaccine 31(Suppl 2):B61–B72. https://doi.org/10.1016/j.vaccine.2012.11.035

Levin MJ et al (2003) Decline in varicella-zoster virus (VZV)-specific cell-mediated immunity with increasing age and boosting with a high-dose VZV vaccine. J Infect Dis 188:1336–1344. https://doi.org/10.1086/379048 (JID30798 [pii])

Lieberman JM et al (2006) The safety and immunogenicity of a quadrivalent measles, mumps, rubella and varicella vaccine in healthy children: a study of manufacturing consistency and persistence of antibody. Pediatr Infect Dis J 25:615–622. https://doi.org/10.1097/01.inf.0000220209.35074.0b00006454-200607000-00010 [pii]

Liko J, Robison SG, Cieslak PR (2013) Priming with whole-cell versus acellular pertussis vaccine. N Engl J Med 368:581–582. https://doi.org/10.1056/nejmc1212006

Lopez AS et al (2006) One dose of varicella vaccine does not prevent school outbreaks: is it time for a second dose? Pediatrics 117:e1070–1077. https://doi.org/10.1542/peds.2005-2085 (117/6/e1070 [pii])

Marin M, Guris D, Chaves SS, Schmid S, Seward JF, Advisory Committee on Immunization Practices CfDC, Prevention (2007) Prevention of varicella: recommendations of the Advisory Committee on Immunization Practices (ACIP). MMWR Recomm Rep 56:1–40

Marin M, Marlow M, Moore KL, Patel M (2018) Recommendation of the Advisory Committee on Immunization Practices for use of a third dose of mumps virus-containing vaccine in persons at increased risk for mumps during an outbreak. MMWR Morb Mortal Wkly Rep 67:33–38. https://doi.org/10.15585/mmwr.mm6701a7

Medical Research Council (1956) Vaccination against whooping-cough; relation between protection in children and results of laboratory tests; a report to the Whooping-cough Immunization Committee of the Medical Research Council and to the medical officers of health for Cardiff, Leeds, Leyton, Manchester, Middlesex, Oxford, Poole, Tottenham, Walthamstow, and Wembley. Br Med J 2:454–462

Miller AE (1980) Selective decline in cellular immune response to varicella-zoster in the elderly. Neurology 30:582–587

Misegades LK, Winter K, Harriman K, Talarico J, Messonnier NE, Clark TA, Martin SW (2012) Association of childhood pertussis with receipt of 5 doses of pertussis vaccine by time since last vaccine dose, California, 2010. JAMA 308:2126–2132. https://doi.org/10.1001/jama.2012. 14939

Morrison VA et al (2015) Long-term persistence of zoster vaccine efficacy. Clin Infect Dis 60:900–909. https://doi.org/10.1093/cid/ciu918ciu918 [pii]

Morse JL (1913) Whooping-cough: a plea for more efficient public regulations relative to the control of this most serious and fatal disease. JAMA 60:1677–1680

Murphy TV et al (2001) Intussusception among infants given an oral rotavirus vaccine. N Engl J Med 344:564–572. https://doi.org/10.1056/NEJM200102223440804

Murray NE, Quam MB, Wilder-Smith A (2013) Epidemiology of dengue: past, present and future prospects. Clin Epidemiol 5:299–309. https://doi.org/10.2147/clep.s34440

Nair H et al (2010) Global burden of acute lower respiratory infections due to respiratory syncytial virus in young children: a systematic review and meta-analysis. Lancet 375:1545–1555. https:// doi.org/10.1016/S0140-6736(10)60206-1

Novartis Vaccines and Diagnostics (2015) BEXSERO® (Meningococcal Group B Vaccine) package insert. Cambridge, MA

Oxman MN (1995) Immunization to reduce the frequency and severity of herpes zoster and its complications. Neurology 45:S41–S46

Oxman MN et al (2005) A vaccine to prevent herpes zoster and postherpetic neuralgia in older adults. N Engl J Med 352:2271–2284. https://doi.org/10.1056/nejmoa051016 (352/22/2271 [pii])

Paavonen J et al (2007) Efficacy of a prophylactic adjuvanted bivalent L1 virus-like-particle vaccine against infection with human papillomavirus types 16 and 18 in young women: an interim analysis of a phase III double-blind, randomised controlled trial. Lancet 369:2161–2170

Pedraz C, Carbonell-Estrany X, Figueras-Aloy J, Quero J, Group IS (2003) Effect of palivizumab prophylaxis in decreasing respiratory syncytial virus hospitalizations in premature infants. Pediatr Infect Dis J 22:823–827. https://doi.org/10.1097/01.inf.0000086403.50417.7c

Peltola H, Kayhty H, Virtanen M, Makela PH (1984) Prevention of Hemophilus influenzae type b bacteremic infections with the capsular polysaccharide vaccine. N Engl J Med 310:1561–1566. https://doi.org/10.1056/NEJM198406143102404

Plosker GL (2012) A/H5N1 prepandemic influenza vaccine (whole virion, vero cell-derived, inactivated) [Vepacel(R)]. Drugs 72:1543–1557. https://doi.org/10.2165/11209650-000000000-00000

Plotkin SA (2014) The pertussis problem Clin Infect Dis 58:830–833. https://doi.org/10.1093/cid/ cit934

Polack FP (2007) Atypical measles and enhanced respiratory syncytial virus disease (ERD) made simple. Pediatr Res 62:111–115. https://doi.org/10.1203/PDR.0b013e3180686ce0

Prevots DR, Sutter RW, Strebel PM, Weibel RE, Cochi SL (1994) Completeness of reporting for paralytic poliomyelitis, United States, 1980 through 1991. Implications for estimating the risk of vaccine-associated disease. Arch Pediatr Adolesc Med 148:479–485

Provost PJ, Hilleman MR (1979) Propagation of human hepatitis A virus in cell culture in vitro. Proc Soc Exp Biol Med 160:213–221

Provost PJ, Hughes JV, Miller WJ, Giesa PA, Banker FS, Emini EA (1986) An inactivated hepatitis A viral vaccine of cell culture origin. J Med Virol 19:23–31

Purcell RH, D'Hondt E, Bradbury R, Emerson SU, Govindarajan S, Binn L (1992) Inactivated hepatitis A vaccine: active and passive immunoprophylaxis in chimpanzees. Vaccine 10(Suppl 1):S148–S151

Rauh LW, Schmidt R (1965) Measles immunization with killed virus vaccine. Serum Antibody Titers and Experience with Exposure to Measles Epidemic. Am J Dis Child 109:232–237

Rendi-Wagner P, Korinek M, Winkler B, Kundi M, Kollaritsch H, Wiedermann U (2007) Persistence of seroprotection 10 years after primary hepatitis A vaccination in an unselected study population. Vaccine 25:927–931. https://doi.org/10.1016/j.vaccine.2006.08.044

Reyes IS, Hsieh DT, Laux LC, Wilfong AA (2011) Alleged cases of vaccine encephalopathy rediagnosed years later as Dravet syndrome. Pediatrics 128:e699–e702. https://doi.org/10. 1542/peds.2010-0887

Richie TL et al (2015) Progress with Plasmodium falciparum sporozoite (PfSPZ)-based malaria vaccines. Vaccine 33:7452–7461. https://doi.org/10.1016/j.vaccine.2015.09.096s0264-410x (15)01386-9 [pii]

Roper M, Wassilak S, Tiwari T, Orenstein W (2013) Tetanus toxoid. In: Plotkin SA, Orenstein WA, Offit PA (eds) Vaccines, 6th edn. Saunders/Elsevier, Philadelphia, pp 746–772

RTS-S Clinical Trials Partnership (2015) Efficacy and safety of RTS,S/AS01 malaria vaccine with or without a booster dose in infants and children in Africa: final results of a phase 3, individually randomised, controlled trial. Lancet 386:31–45. https://doi.org/10.1016/s0140-6736(15)60721-8s0140-6736(15)60721-8 [pii]

Sabchareon A et al. (2012) Protective efficacy of the recombinant, live-attenuated, CYD tetravalent dengue vaccine in Thai schoolchildren: a randomised, controlled phase 2b trial. Lancet 380:1559–1567. https://doi.org/10.1016/s0140-6736(12)61428-7 (S0140-6736(12)61428-7 [pii])

Safaeian M et al (2013) Durable antibody responses following one dose of the bivalent human papillomavirus L1 virus-like particle vaccine in the Costa Rica vaccine trial. Cancer Prev Res (Phila) 6:1242–1250. https://doi.org/10.1158/1940-6207.capr-13-0203

Sako W (1947) Studies on pertussis immunization. J Pediatr 30:29–40

Sanofi Pasteur (2017) Dengue vaccine registered in 19 countries. http://dengue.info/dengue-vaccine-registered-in-19-countries/. Accessed 12 Feb 2018

Santosham M, Reid R, Letson GW, Wolff MC, Siber G (1990) Passive immunization for infection with Haemophilus influenzae type b. Pediatrics 85:662–666

Sato H, Sato Y (1990) Protective activities in mice of monoclonal antibodies against pertussis toxin. Infect Immun 58:3369–3374

Schiller JT, Hidesheim A (2000) Developing HPV virus-like particle vaccines to prevent cervical cancer: a progress report. J Clin Virol 19:67–74

Schiller JT, Lowy DR (1996) Papillomavirus-like particles and HPV vaccine development. Semin Cancer Biol 7:373–382. https://doi.org/10.1006/scbi.1996.0046

Schmader KE et al (2012) Persistence of the efficacy of zoster vaccine in the shingles prevention study and the short-term persistence substudy. Clin Infect Dis 55:1320–1328. https://doi.org/10.1093/cid/cis638cis638 [pii]

Schneerson R, Rodrigues LP, Parke JC Jr, Robbins JB (1971) Immunity to disease caused by Hemophilus influenzae type b. II. Specificity and some biologic characteristics of "natural," infection-acquired, and immunization-induced antibodies to the capsular polysaccharide of Hemophilus influenzae type b. J Immunol 107:1081–1089

Schwarz TF, Leo O (2008) Immune response to human papillomavirus after prophylactic vaccination with AS04-adjuvanted HPV-16/18 vaccine: improving upon nature. Gynecol Oncol 110:S1–S10. https://doi.org/10.1016/j.ygyno.2008.05.036

Seder RA et al (2013) Protection against malaria by intravenous immunization with a nonreplicating sporozoite vaccine. Science 341:1359–1365. https://doi.org/10.1126/science.1241800 [pii]

Sheridan SL, Ware RS, Grimwood K, Lambert SB (2012) Number and order of whole cell pertussis vaccines in infancy and disease protection. JAMA 308:454–456. https://doi.org/10.1001/jama.2012.6364

Shorvon S, Berg A (2008) Pertussis vaccination and epilepsy—an erratic history, new research and the mismatch between science and social policy. Epilepsia 49:219–225. https://doi.org/10.1111/j.1528-1167.2007.01478.x

Simmons CP (2015) A candidate dengue vaccine walks a tightrope. N Engl J Med 373:1263–1264. https://doi.org/10.1056/NEJMe1509442

Slifka MK, Amanna I (2014) How advances in immunology provide insight into improving vaccine efficacy. Vaccine 32:2948–2957. https://doi.org/10.1016/j.vaccine.2014.03.078s0264-410x(14)00459-9 [pii]

Stanaway JD et al (2016) The global burden of dengue: an analysis from the Global Burden of Disease Study 2013. Lancet Infect Dis https://doi.org/10.1016/s1473-3099(16)00026-8

Stokes J Jr, Neefe JR (1945) The prevention and attenuation of infectious hepatitis by gamma globulin. JAMA 127:144–145

Stokley S et al (2014) Human papillomavirus vaccination coverage among adolescents, 2007–2013, and postlicensure vaccine safety monitoring, 2006–2014—United States. MMWR Morb Mortal Wkly Rep 63:620–624

Storsaeter J, Hallander HO, Gustafsson L, Olin P (1998) Levels of anti-pertussis antibodies related to protection after household exposure to *Bordetella pertussis*. Vaccine 16:1907–1916

Strebel PM, Papania MJ, Fiebelkorn AP, Halsey NA (2013) Measles vaccines. In: Plotkin SA, Orenstein WA, Offit PA (eds) vaccines, 6th edn. Saunders/Elsevier, Philadelphia, pp 352–388

Tejedor JC et al (2012) Five-year antibody persistence and safety following a booster dose of combined Haemophilus influenzae type b-Neisseria meningitidis serogroup C-tetanus toxoid conjugate vaccine. Pediatr Infect Dis J 31:1074–1077. https://doi.org/10.1097/INF.0b013e318269433a

The FUTURE II Study Group (2007) Quadrivalent vaccine against human papillomavirus to prevent high-grade cervical lesions. N Engl J Med 356:1915–1927

The IMpact RSV Study Group (1998) Palivizumab, a humanized respiratory syncytial virus monoclonal antibody, reduces hospitalization from respiratory syncytial virus infection in high-risk infants. Pediatrics 102:531–537

Theeten H, Van Herck K, Van Der Meeren O, Crasta P, Van Damme P, Hens N (2015) Long-term antibody persistence after vaccination with a 2-dose Havrix (inactivated hepatitis A vaccine): 20 years of observed data, and long-term model-based predictions. Vaccine 33:5723–5727. https://doi.org/10.1016/j.vaccine.2015.07.008

Tiwari T, Wharton M (2013) Diphtheria toxoid. In: Plotkin SA, Orenstein WA, Offit PA (eds) Vaccines, 6th edn. Saunders/Elsevier, Philadelphia, pp 153–166

Tseng HF et al (2016) Declining effectiveness of herpes zoster vaccine in adults aged >/=60. Years J Infect Dis. https://doi.org/10.1093/infdis/jiw047 (jiw047 [pii])

Van Damme P, Thoelen S, Cramm M, De Groote K, Safary A, Meheus A (1994) Inactivated hepatitis A vaccine: reactogenicity, immunogenicity, and long-term antibody persistence. J Med Virol 44:446–451

Vaughn DW et al (2000) Dengue viremia titer, antibody response pattern, and virus serotype correlate with disease severity J Infect Dis 181:2–9. https://doi.org/10.1086/315215 (JID990867 [pii])

Vazquez M, LaRussa PS, Gershon AA, Steinberg SP, Freudigman K, Shapiro ED (2001) The effectiveness of the varicella vaccine in clinical practice. N Engl J Med 344:955–960. https://doi.org/10.1056/nejm200103293441302 (MJBA-441302 [pii])

Villar L et al (2015) Efficacy of a tetravalent dengue vaccine in children in Latin America. N Engl J Med 372:113–123. https://doi.org/10.1056/NEJMoa1411037

von Behring E, Kitasato S (1890) Ueber das zustandekommen der diphtherie-immunitat und der tetanus-immunitat bei thieren (On the realization of immunity in diphtheria and tetanus in animals). Dtsch Med Wochenschr 16:1113–1114

Walton LR, Orenstein WA, Pickering LK (2015) Lessons learned from making and implementing vaccine recommendations in the U.S. Vaccine 33(Suppl 4):D78–D82. https://doi.org/10.1016/j.vaccine.2015.09.036

Ward JI, Broome CV, Harrison LH, Shinefield H, Black S (1988) Haemophilus influenzae type b vaccines: lessons for the future. Pediatrics 81:886–893

Webster DP, Farrar J, Rowland-Jones S (2009) Progress towards a dengue vaccine. Lancet Infect Dis 9:678–687. https://doi.org/10.1016/s1473-3099(09)70254-3 (S1473-3099(09)70254-3 [pii])

Werzberger A et al (1992) A controlled trial of a formalin-inactivated hepatitis A vaccine in healthy children. N Engl J Med 327:453–457. https://doi.org/10.1056/NEJM199208133270702

Wiedermann G, Kundi M, Ambrosch F, Safary A, D'Hondt E, Delem A (1997) Inactivated hepatitis A vaccine: long-term antibody persistence. Vaccine 15:612–615

Williamson ED, Oyston PCF (2013) Plague vaccines. In: Plotkin SA, Orenstein WA, Offit PA (eds) Vaccines, 5th edn. Saunders/Elsevier, Philadelphia, pp 493–503

Wilson A, Sharp M, Koropchak CM, Ting SF, Arvin AM (1992) Subclinical varicella-zoster virus viremia, herpes zoster, and T lymphocyte immunity to varicella-zoster viral antigens after bone marrow transplantation. J Infect Dis 165:119–126

Wolters KL, Dehmel H (1942) [Final investigations into tetanus prevention by active immunization] Zeitschrift für hygiene und infektions krankheiten medizinische, mikrobiologie und viralogie 124:326–332

World Health Organization (2014) Global control and regional elimination of measles, 2000–2012. Wkly Epidemiol Rec 89:45–52

Young MK, Nimmo GR, Cripps AW, Jones MA (2014) Post-exposure passive immunisation for preventing measles. Cochrane Database Syst Rev 4:CD010056 https://doi.org/10.1002/14651858.cd010056.pub2

Zhou F et al (2005) Economic evaluation of the 7-vaccine routine childhood immunization schedule in the United States, 2001. Arch Pediatr Adolesc Med 159:1136–1144. https://doi.org/10.1001/archpedi.159.12.1136 (159/12/1136 [pii])

Zingher A (1924) Convalescent whole blood plasma and serum in prophylaxis of measles. JAMA 82:1180–1187

Broadly Neutralizing Antibodies to Highly Antigenically Variable Viruses as Templates for Vaccine Design

Matthias G. Pauthner and Lars Hangartner

Contents

Abstract Development of vaccines to highly variable viruses such as Human Immunodeficiency Virus and influenza A viruses faces multiple challenges. In this article, these challenges are described and reverse vaccinology approaches to generate universal vaccines against both pathogens are laid out and compared.

M. G. Pauthner · L. Hangartner (✉)
The Scripps Research Institute, 10550 North Torrey Pines Road, La Jolla, CA 92037, USA
e-mail: lhangart@scripps.edu

M. G. Pauthner
e-mail: pauthner@scripps.edu

Current Topics in Microbiology and Immunology (2020) 428: 31–87
https://doi.org/10.1007/82_2020_221

1 Introduction: Broadly Neutralizing Antibodies— Challenges and Solutions

1.1 Requirements for Successful Vaccines to Viral Pathogens

Successful vaccines to viral pathogens largely rely on the induction of functional antibody responses to effectively protect the host from infection (Plotkin 2010, 2013). While antibodies can aid host protection through a multitude of Fc-mediated functions, e.g., via antibody-dependent cellular cytotoxicity (ADCC), the most important correlate of antibody-mediated protection from viral pathogens is the ability to inhibit productive infection of host cells, or direct viral neutralization. Neutralization is typically measured by incubating antibodies or polyclonal sera of interest with live or pseudo-typed virus before transfer onto an appropriate reporter cell line to indicate inhibition of infection (Li et al. 2005). In case of influenza, although neutralization assays have become more widely used, hemagglutination inhibition assays (HAI) are still quite common to determine antibody-mediated protection in humans; HAI titers of or above 1:40 are considered as protective (Coudeville et al. 2010). However, HAI assays only detect inhibition of receptor binding and will not quantify neutralizing antibodies that prevent host cell infection by means other than blocking receptor binding (Bachmann et al. 1999; Wyrzucki et al. 2014; Maurer et al. 2018). Direct neutralization is considered a consequence of high-affinity binding of neutralizing antibodies (nAbs) to an exposed epitope on the infectious virus particle. While the exact mechanism(s) of neutralization are frequently unknown for a given nAb, the most common and best understood mechanisms of direct neutralization are (i) steric obstruction of viral entry (attachment or fusion) into the host cell and (ii) allosteric mechanisms, e.g., by hindering conformational changes in the viral receptor protein and thus preventing the formation of the viral fusion spike (Julien et al. 2013b). Passive transfer studies of HIV nAbs with intact or impaired Fc-effector domains in vivo further solidified that direct neutralization is sufficient for host protection, but can be enhanced by Fc-mediated immune responses to both free virus or infected cells (e.g., ADCC) (Hessell et al. 2007).

To achieve sterilizing immunity from infection, vaccines need to be able to generate memory responses of potent nAbs upon immunization, comprised of both circulating levels of nAbs, which are continuously produced by long-lived plasma cells in the bone marrow and corresponding memory B cells, which can rapidly morph into plasmablasts and secrete nAbs into the bloodstream upon antigen contact. It is important to note that different vaccines may rely on different memory components for success—e.g., to prevent HIV infection, readily available, circulating nAbs are likely required as infection of a single, or small number of cells is deemed sufficient to establish latent infection. Although antibodies are of importance, e.g., by blunting viremia, clearance of acute viral infections is often the result of the cytotoxic T lymphocytes (CTLs) (Borrow et al. 1994). However, preexisting antibodies are the only means to prevent (re-)infection and can provide immunity by inactivating pathogens before they are able to infect target cells, also referred to as sterilizing immunity. Moreover, maternal antibodies can extend maternal immunity to the offspring.

Traditional vaccination approaches with live-attenuated or inactivated whole virus particles, as well as multimerized protein subunits, so-called virus like particles (VLPs), are in fact very effective at inducing long-lasting, protective immune responses. Advances in vaccine research enabled the global eradication of smallpox (Fenner 1993), the near-complete eradication of polio (Greene et al. 2019), as well as sharp reductions in the incidence of many common viral diseases, such as measles, mumps, rubella and hepatitis A and B in most countries with vaccination rates high enough to sustain herd immunity.

1.2 Highly Antigenically Variable Viruses

However, a subset of pathogens defy traditional vaccination approaches due to their extraordinarily high rates of genetic variability paired with a variety of immune-evasive measures. These pathogens include bacteria (e.g., Streptococcus pneumoniae), protozoa like Plasmodium falciparum and Trypanosoma brucei, the causative agents of malaria and sleeping sickness, respectively, and most notably RNA viruses such as HIV, influenza virus and hepatitis C virus (HCV) (Burton et al. 2012).

The driving force behind the enormous genetic variability of RNA viruses is the low fidelity of viral RNA-dependent polymerases. Reported mutation rates for riboviruses (and some single-stranded DNA viruses) are in the range of 10^{-4}–10^{-6} per nucleotide site per infection cycle, which is strongly elevated compared to DNA viruses that range from 10^{-6} to 10^{-8} errors per nucleotide per infection cycle (Sanjuán et al. 2010). Retroviruses have comparable mutation rates to RNA viruses, averaging $\sim 2.0 \times 10^{-5}$ errors per nucleotide per infection cycle. By comparison, the human genome was estimated to replicate with only $\sim 1.1 \times 10^{-8}$ errors per nucleotide per haploid genome (Roach et al. 2010). Given the vast proliferative capacity of RNA viruses in natural infection paired with their small genome size (typically 5–15 kbp), the high error rate could theoretically generate every possible point mutation and many double mutations during each round of replication in a given host (Lauring et al. 2013). However, it is important to note that high genetic variability also causes a substantial amount of fitness-reducing and lethal mutations. For Vesicular Stomatitis Virus (VSV), for example, 70% of random mutations were found to be either deleterious or lethal, while only 5–10% showed fitness enhancement, the remainder being neutral (Sanjuán et al. 2004). Purifying selection thus reduces the antigenic diversity presented to the host immune system, which is typically derived from viral quasi-species with highly competitive replicative fitness (Wood et al. 2009). In the case of HIV, high selection pressures associated with sexual transmission further reduce viral diversity immediately following infection, often to a single viral isolate, so-called transmitted founder viruses (Keele et al. 2008; Parrish et al. 2015).

Because of the relatively long response time of approximately 7 days that the humoral immune system requires to mount an antibody response to a novel target, the rapid mutation rates and large intra-host population sizes of RNA viruses pose a tremendous threat. Strain-specific antibodies targeting variable epitopes on surface-exposed viral proteins do apply selection pressure (Laver et al. 1979a, b; Both et al. 1983), at best selecting for escape mutants with reduced replicative fitness, but inevitably becoming obsolete upon viral escape. However, by necessity, all viruses also comprise epitopes that are essential to their survival and therefore generally cannot be altered without some fitness penalty. These essential, conserved epitopes commonly include receptor-binding sites, required for host cell tethering and binding, as well as fusion proteins, enabling membrane fusion with the host cell. Thus, high antigenic variability is not necessarily problematic, as long as conserved epitopes exist that can be readily targeted by the immune system. A great example for this is measles virus, which is highly genetically diversified into eight clades and numerous subtypes and has a mutation rate of 4.4×10^{-5} errors per nucleotide per replication cycle, which is an order of magnitude higher than influenza A virus (Sanjuán et al. 2010), and comparable to the average mutation rate per replication cycle measured for HIV ($\sim 3.6 \times 10^{-5}$) (Rawson et al. 2016). However, unlike HIV and influenza virus, measles virus has a single serotype and is reliably neutralized by vaccine-induced nAbs targeting epitopes surrounding its conserved receptor-binding site on measles hemagglutinin (Tahara et al. 2012).

Consequently, a productive way to think about the level of difficulty associated with vaccine development for a given viral pathogen is to ask how immunodominant its conserved, life-cycle-dependent epitopes are. As discussed in-depth in Chap. 4, immunodominance can be defined as the likelihood of a successful immune response against a given epitope. This in turn is a measure of the affinity of the epitope to the closest germline B cell receptor (BCR), the abundance of appropriate precursor B cells and the relative frequency of the target epitope on the viral surface, which affects binding avidity and as a consequence BCR cross-linking and activation. It follows that prominent viral evasion strategies aim at lowering the immunodominance of essential epitopes. A common strategy deployed by both influenza virus and HIV is the camouflaging of conserved epitopes with N-linked glycans (Sok et al. 2016b; Wu and Wilson 2017). N-linked glycans are sterically obstructive, typically associated with immunologic self (Scanlan et al. 2007) and are easily added and removed by mutations of the $N \times T/N \times S$ glycosylation signals, where x represents any amino acid other than proline.

In this context, two classes of viruses can be discerned, based on their requirement to persist in the presence of immune countermeasures: evasion strong and evasion lite viruses (Hangartner et al. 2006; Burton 2017). While some evasion lite viruses display high antigenic diversity, their life cycle allows them to move on to a new host quickly, thus reducing the fitness advantage of sophisticated and burdening measures of immune evasion (e.g., measles and polio). On the other hand, viruses that need to thrive in antibody-rich environments during their infection cycle, e.g., HIV, influenza virus and HCV, have developed refined strategies of immune evasion, including camouflaging and obstructing essential epitopes, which poses a substantial problem for vaccine design. As traditional vaccination approaches consistently failed to create universal vaccines for these pathogens, entirely new and more focused strategies are needed to tackle evasion strong viruses, most prominently HIV and influenza virus, on which we will focus for the remainder of this review.

1.3 Broadly Neutralizing Antibodies and Reverse Vaccinology

In the case of HIV, a small subset of 10–30% of long-term infected individuals manage to circumvent viral evasion measures and develop some level of nAbs that mostly derive their binding energy from contacts with relatively conserved surface residues (McCoy and McKnight 2017). Some of these antibodies very potently neutralize a wide range of HIV isolates (Sok et al. 2014b), as high as 98% of global isolates (Huang et al. 2016), or, in the case of influenza A, potently neutralize all hemagglutinin (HA) subtypes across both serologic groups (Corti et al. 2011; Dreyfus et al. 2012; Wyrzucki et al. 2015). These antibodies have been dubbed broadly neutralizing antibodies (bnAbs), due to their extraordinary breadth in neutralization, as compared to strain-specific nAbs that target variable, non-essential epitopes.

To overcome the structural restrains that HIV and influenza virus have evolved over time, bnAbs against these pathogens show a variety of unusual genetic features that have made their re-elicitation through vaccination challenging, although tremendous progress has been made in recent years (Burton and Hangartner 2016; Sautto et al. 2018). The discovery and distinct features of bnAbs will be discussed in the following section, followed by a brief review of correlates of protection from HIV infection that have been identified in experimental HIV infection models and clinical vaccine trials.

There are several vaccination strategies that seek to induce bnAbs, which will be briefly discussed at the end of this chapter. The basic concept underlying all of these strategies is a detailed molecular understanding of the broadly neutralizing epitopes and their corresponding bnAbs, which is than leveraged using protein design to develop rationally designed immunogens. This concept (Fig. 1) is referred to as 'reverse vaccinology 2.0' as it starts with the neutralizing antibody and aims to reverse engineer a vaccine candidate that is complementary in shape, in a reverse of the traditional flow of vaccine design (Burton 2002, 2017). The following sections

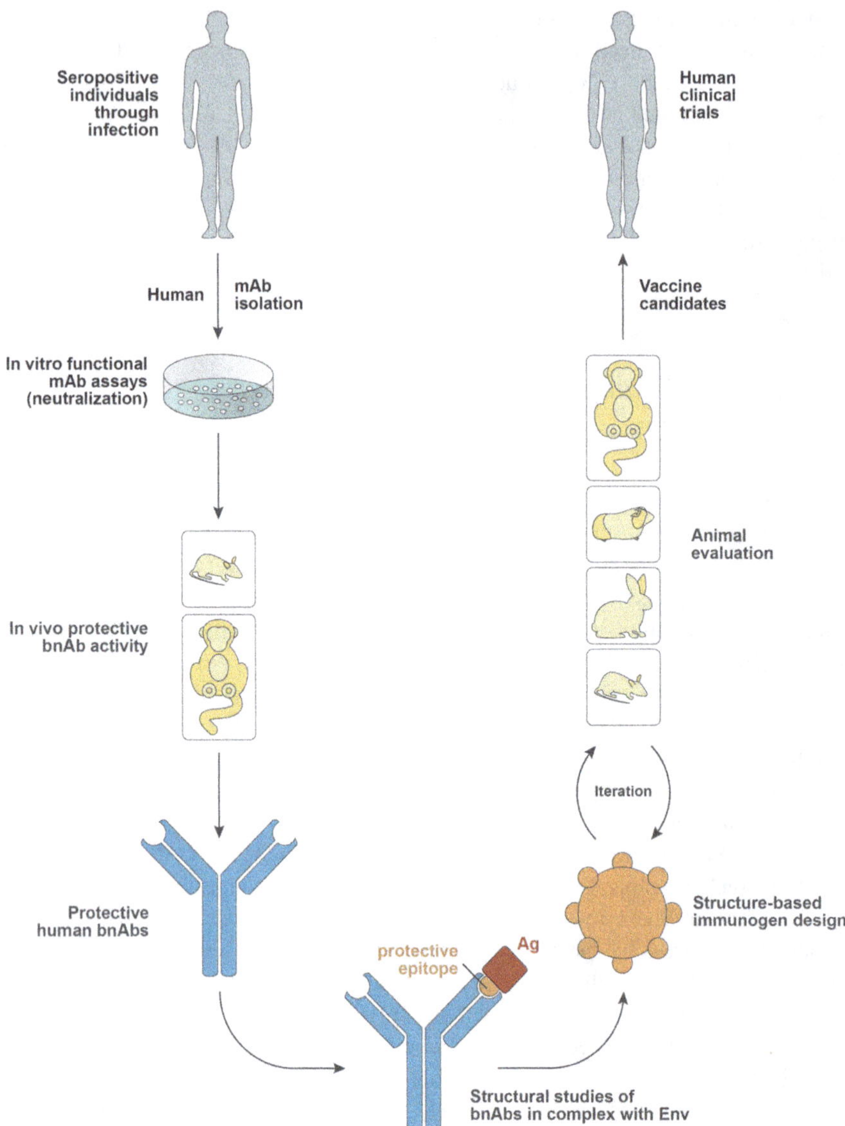

Fig. 1 The reverse vaccinology 2.0 workflow from mAb isolation to iterative vaccine design and clinical trials

will summarize the known details of bnAbs targeting HIV and influenza virus, respectively, followed by a short discussion of how these findings are currently being utilized to inform vaccine design.

2 Broadly Neutralizing Antibodies to HIV

2.1 The Envelope Trimer, the Sole Target of bnAbs to HIV

Over the last two decades, the isolation of broadly HIV neutralizing antibodies has grown exponentially, which has paralleled a gradual shift in the HIV vaccine field from a strong focus on CD8$^+$ T cell induction to strategies aimed at the elicitation of bnAbs (Burton and Hangartner 2016). The sole target of neutralizing antibodies to HIV is the exposed envelope (Env) protein, which forms the so-called envelope spike above the viral membrane. Envelope's function is to bind CD4 as well as CCR5 or CXCR4 as co-receptors on the surface of immune cells, which then initiates membrane fusion and viral entry (Dalgleish et al. 1984). Membrane-coated HIV virions are approximately 110 nm in diameter, but display an average of only 8–14 Env spikes on the virion surface (Zhu et al. 2006; Zanetti et al. 2006), which represents a rather low spike density compared to other enveloped RNA viruses, such as influenza virus (Fig. 2). Interestingly, the surface spike density is regulated by the intracellular C-terminal (Ct) domain of Env, deletion of which increases the spike density by almost an order of magnitude to 73–88 per virion (Zanetti et al. 2006). A recent study suggests that only 2–3 spikes per virion need to be engaged

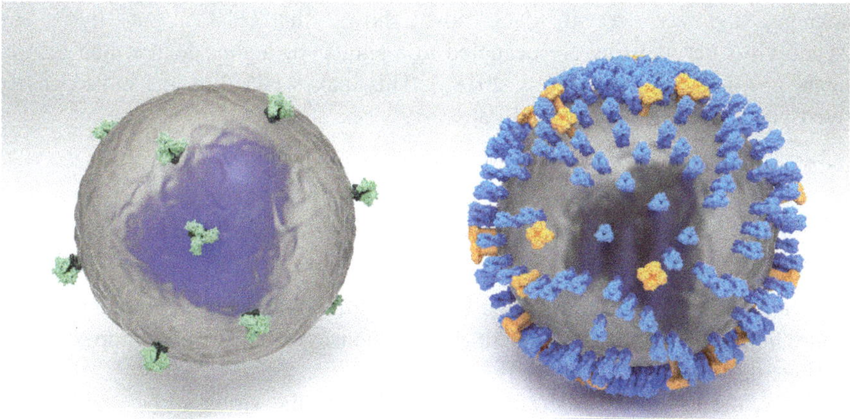

Fig. 2 Comparison of the virion structures of HIV-1 (left) and influenza A (right). The virions of HIV and influenza A virus differ considerably in the number and density of viral spikes. While only around 14 HIV envelope spikes can be found on HIV (Zhu et al. 2006), there are an estimated 290–350 HA, and 35–50 NA spikes present on influenza A virions (Harris et al. 2006). These different spike densities are likely to influenza epitope accessibility: While HIV trimers are fully accessible from all sides, access to lateral and basal epitopes on influenza A virions may be restricted due to the dense packing. Virion models were constructed to scale in Blender 2.78 using the surface meshes for HIV Env (PDB 2HTY), and H1 HA trimers (PDB 1RD89). The NA tetramer was modeled using the structure for the globular N1 head (PDB 2HTY) and stalk modeled from known dimensions

to facilitate viral entry; the exact number was found to be strain dependent and varied from 1 to 7 infectious spikes per virion (Brandenberg et al. 2015).

Envelope is translated as a single heavily glycosylated protein, named gp160. As gp160 is translated in the ER and shuttled through the Golgi apparatus on its way to the plasma membrane, two important post-translational modifications take place that are important for the overall structure and immunogen design considerations. Firstly, gp160 is cleaved into its major building blocks gp120, which is comprised of five variable loops (V1–V5) flanked by constant regions (C1–C5) and contains the receptor-binding site, as well as gp41, which contains the fusion machinery. The cleavage is mediated by the host protease furin, which recognizes a conserved sequence of 4–6 lysines in the gp160 peptide sequence (Decroly et al. 1994). On the cell surface, gp120 and gp41 then form what is commonly described as a trimer of hetero-dimers, with three subunits of membrane-anchored gp41 forming the trimeric pedestal upon which a second, non-covalently attached trimer of gp120 subunits rests. Secondly, the $\sim$25 potential N-linked glycosylation sites (PNGS) per gp120 and 4 PNGS sites per gp41 subunit (Korber et al. 2001) become glycosylated during their interactions with glycosyltransferases and glycosidases in the ER and Golgi apparatus (Doores et al. 2010). Recent studies using novel mass spectrometry approaches (Cao et al. 2018a) have shown that PNGS on recombinant gp120 (Struwe et al. 2017) as well as stabilized gp140 trimers (Cao et al. 2017), are in fact almost fully occupied. While similar data sets for virion-derived gp120 existed (Panico et al. 2016; Pritchard et al. 2015a), glycan occupancy analysis was lacking. However, recent data now shows that PNGS on live- and pseudo-virus-derived Env are occupied to a similar or higher degree than recombinant gp140 trimers (Cao et al. 2018b). This makes HIV Env the most densely glycosylated protein described to date.

2.2 Immune-Evasive Features of the Native Env Trimer

There are three main features of the HIV Env spike that deter the immune system from efficiently mounting neutralizing Ab responses to conserved epitopes cluster on Env.

Firstly, the low density of native Env spikes on the virion surface (Zhu et al. 2006) is a major immune-evasive mechanism of envelope, as the scarce spacing of spikes disfavors B cell receptor cross-linking and reduces binding avidity and consequently B cell activation (Dintzis et al. 1976).

Secondly, many conserved, broadly neutralizing epitopes are conformational and only displayed correctly in the context of the native Env trimer. The non-covalent attachment of gp120 promoters to the gp41 base, however, renders the Env spike inherently labile or meta-stable. Consequently, non-native Env trimers with missing or splayed-open gp120 crowns, as well as shed gp120, are presented to the immune system alongside native trimers, thus presenting a multitude of non-native, non-neutralizing epitopes that dilute the frequency of desirable (broadly) neutralizing epitopes (Poignard et al. 2003; Moore et al. 2006; Crooks et al. 2011). Another immune-evasive aspect of the trimeric spike is the steric occlusion of essential epitopes, e.g., the CD4 binding site, in poorly accessible locations between promoters, proximal to the C3 symmetry axis of the trimer (Julien et al. 2013a; Lyumkis et al. 2013; Pancera et al. 2014).

The third, and likely most significant immune counter-measure that HIV commands, is the dense layer of oligomannose, hybrid- and complex-type glycans covering the protein surface, which is commonly referred to as the glycan shield (Wei et al. 2003). Reasons for the shielding property include: (i) due to the hydrophilic nature of glycans, protein–glycan-binding interactions yield lower free-binding energy per buried interface area than protein–protein interactions (Liang et al. 2007). This energy difference means that antibodies targeting glycan or proteoglycan epitopes must bury large amounts of surface area, i.e., increasing their binding footprint, to accumulate enough free-binding energy needed for high-affinity interactions. A larger binding footprint, in turn, renders nAbs more susceptible to viral escape, especially when conserved epitopes are flanked by highly variable areas. Most glycan-binding bnAbs to HIV also make additional contacts with the underlying peptide surface, which often requires unusual genetic features like very long HCDR3s to reach through the dense glycan layer. The requirement of these features decreases the frequency of such bnAbs naturally occurring, as they typically require extensive somatic hypermutation (SHM) during affinity maturation in germinal center (GC) reactions to evolve. (ii) The viral genome has evolved to facilitate shifts or complete loss and addition of PNGS with extraordinary ease (Coss et al. 2016). In many cases, single point mutations suffice to alter the glycan shield composition, thus enabling rapid viral escape from nAbs, either by shifting or deleting bnAb-bound glycans (MacLeod et al. 2016) (Moore et al. 2012) or by obscuring bnAb epitopes via glycan addition (Sok et al. 2016b). Interestingly, glycan-site deletion in response to nAb-pressure sometimes increases the accessibility of otherwise concealed essential epitopes, like the CD4 binding site or the V3-glycan area, paving the way for subsequent bnAb development (Moore et al. 2012; Gao et al. 2014). (iii) Finally, N-linked glycosylation is a common post-translational modification of mammalian proteins and as such associated with immunologic self. This, at least theoretically, reduces the pool of suitable germline BCRs that are not counter-selected during B cell maturation. However, the frequency and density of N- and O-linked glycosylation sites in the human proteome is very small compared to viral receptor proteins, which has important implications for glycan processing. Of over 20,500 structurally characterized proteins in the PDB, less than 10% are predicted to have N-linked glycosylation sites, and of those proteins bearing PNGS sites, more than 80% only comprise a single PNGS site (Li et al. 2016). Mass spectrometry analysis of the human sperm proteome similarly found only 10% of glycosylated proteins to bear more than three glycan sites (Wang et al. 2013). Low glycan density and comparably slow physiological protein production rates increase glycan accessibility and interaction with the ER and Golgi glycosylation machinery and thus allow for complete processing from oligomannose-type toward complex-type glycans on most human proteins. In the case of Env, however, rapid parallel expression of gp160, extraordinary density of glycosylation sites and structural constraints arising from the trimer structure diminish access and exposure to the glycosylation machinery, thus strongly elevating high-mannose-type glycan frequencies (Doores et al. 2010; Bonomelli et al. 2011; Pritchard et al. 2015a, b). High-mannose-type glycans especially cluster in a relatively conserved area of gp120 dubbed the high-mannose patch, surrounding the glycan at asparagine 332 (N332), as well as the structurally inaccessible areas that result from trimerization of gp120 and gp41 (Doores et al. 2010; Pritchard et al. 2015b). It follows that production cell line, Env transfection ratios, efficient gp160 cleavage and especially monomeric as compared to trimeric protein designs have a notable impact on the glycoforms composition of recombinantly generated immunogens and viruses (Bonomelli et al. 2011; Cao et al. 2017, 2018b).

In summary, although dense glycosylation of Env is a strong deterrent to effective immune responses to HIV, the unusually high density of oligomannose-type glycans on Env also provides the immunological basis for bnAb recognition of glycan-dependent epitopes, largely in the absence of autoreactivity to N-glycosylated human proteins.

2.3 Discovery and Isolation of bnAbs to HIV

The discovery of novel HIV bnAbs happened in two distinct waves, which reflected the technical possibilities at the time, as well as access to well characterized and large cohorts of long-term infected individuals. The first bnAbs to HIV, b12 and 2F5 were isolated in the early 1990s in the laboratories of Dennis Burton and Hermann Katinger. b12 was isolated using phage display, which allowed for the screening of large libraries but did not maintain the original heavy and light chain pairing (Burton et al. 1991, 1994; Barbas et al. 1992). 2F5 was isolated by immortalizing B cells through hybridoma fusion, which was fairly inefficient and limited in throughput (Buchacher et al. 1994; Conley et al. 1994). In the coming years, two more important early bnAbs were described, 2G12 and 4E10 (Buchacher et al. 1994; Zwick et al. 2001; Stiegler et al. 2001). These antibodies showed neutralization breadth of 32–98% on global isolates, respectively, albeit with low median IC_{50}s of above 1 µg/ml (Binley et al. 2004; Walker et al. 2009). However, these antibodies were shown to protect in passive transfer studies in nonhuman primates (NHPs), which generated new hope for antibody-based HIV vaccines (Parren et al. 2001; Hessel et al. 2009, 2010).

The advent of the second generation of HIV bnAbs began with the isolation of PG9 and PG16 by direct neutralization screening of the culture supernatants of 30,000 B cells, derived from a long-term infected donor of the IAVI Protocol G cohort (Walker et al. 2009). What enabled the isolation of these bnAbs was both a rigorous screening and ranking of 1800 Protocol G samples on a cross-clade 6-virus panel (Simek et al. 2009), paired with direct B cell stimulation and culture as a new high-throughput screening technology. B cell cultures supernatants were both tested for binding to HIV Env subunits gp120 and gp41 and for neutralization against isolates SF162 and JR-CSF. Remarkably, PG9 and PG16 were later shown to target a quaternary apex epitope of the trimeric HIV Env spike (Walker et al. 2009; Julien et al. 2013c) with weak or no binding to monomeric gp120 subunits from most HIV strains. Given the absence of soluble native-like trimer proteins at the time, their isolation was only possible by including neutralization of intact HIV pseudo-virions as readout. PG9 and PG16 showed good neutralization breadth of $\sim 75\%$ on a 162-virus panel, but did so with previously unseen potency of ~ 0.2 µg/ml, which was an order of magnitude higher than the first-generation bnAbs (Walker et al. 2009). In the following years, this approach was expanded and directly led to the isolation of the PGT121, PGT128, PGT135, PGT145 and PGT151 bNAb families from multiple Protocol G donors (Walker et al. 2011; Falkowska et al. 2014). PGT121 and PGT128 again pushed the limit of neutralization potency by an order of magnitude to ~ 0.02 µg/ml, while maintaining neutralization breadth of at least 70% on a panel of 162 global isolates (Walker et al. 2011). Based on the low serum concentrations of PGT121 that were sufficient to protect 3/5 rhesus monkeys in a passive transfer study (mean titer 1:285 or 0.18 µg/ml PGT121) (Moldt et al. 2012), this class of bnAbs reinvigorated the hope for globally effective bNAb-based HIV vaccines (Burton et al. 2012).

In parallel, the notion that only few Ab lineages mediate broad serum reactivity (Walker et al. 2010), together with advances in single-cell sorting and especially single-cell PCR amplification of antibody heavy and light chain genes, enabled a new approach for nAb-isolation based on binding of memory B cells to recombinant Env protein probes (Scheid et al. 2009). The first bnAb isolated using this methodology was VRC01. To isolate HIV-reactive memory B cell from a donor whose serology indicated reactivity against the CD4 binding site, a fluorescently labeled gp120 core protein named RSC3 was developed at the Vaccine Research Center (VRC) and used to sort epitope-specific memory B cells from that donor (Wu et al. 2010). The development of recombinant, soluble HIV trimers, so-called SOSIP trimers named after their two key stabilizing mutations, a disulfide-bond between gp120 and gp41 (SOS) (Binley et al. 2000) and an I595P (IP) mutation (Sanders et al. 2002), enabled the isolation of quaternary bnAbs by single-cell sorting. Using SOSIP trimers of the BG505 isolate that displayed authentic antigenicity resembling the native Env spike (Sanders et al. 2013; Julien et al. 2013a; Lyumkis et al. 2013), Env trimers were used to isolate the trimer-specific PGDM1400 Ab family, which again raised the bar for potency of bnAbs, displaying a median IC_{50} of 0.003 µg/ml on large virus panels (Sok et al. 2014b). Both antigen-specific B cell sorting as well as single-cell culture and stimulation approaches have since greatly enriched the repertoire of HIV bNAb families (McCoy and Burton 2017), which, counting clonally related antibodies, encompasses several hundred bnAbs today.

2.4 *Major Broadly Neutralizing Epitopes on HIV Env*

Although a great variety of bnAbs to HIV has been described in recent years, the vast majority can be classified into five major epitope clusters, which are by definition relatively conserved among global isolates (Fig. 3). From 'top to bottom,' these sites of vulnerability are the trimer apex (V2-apex), the glycan-V3 region, the

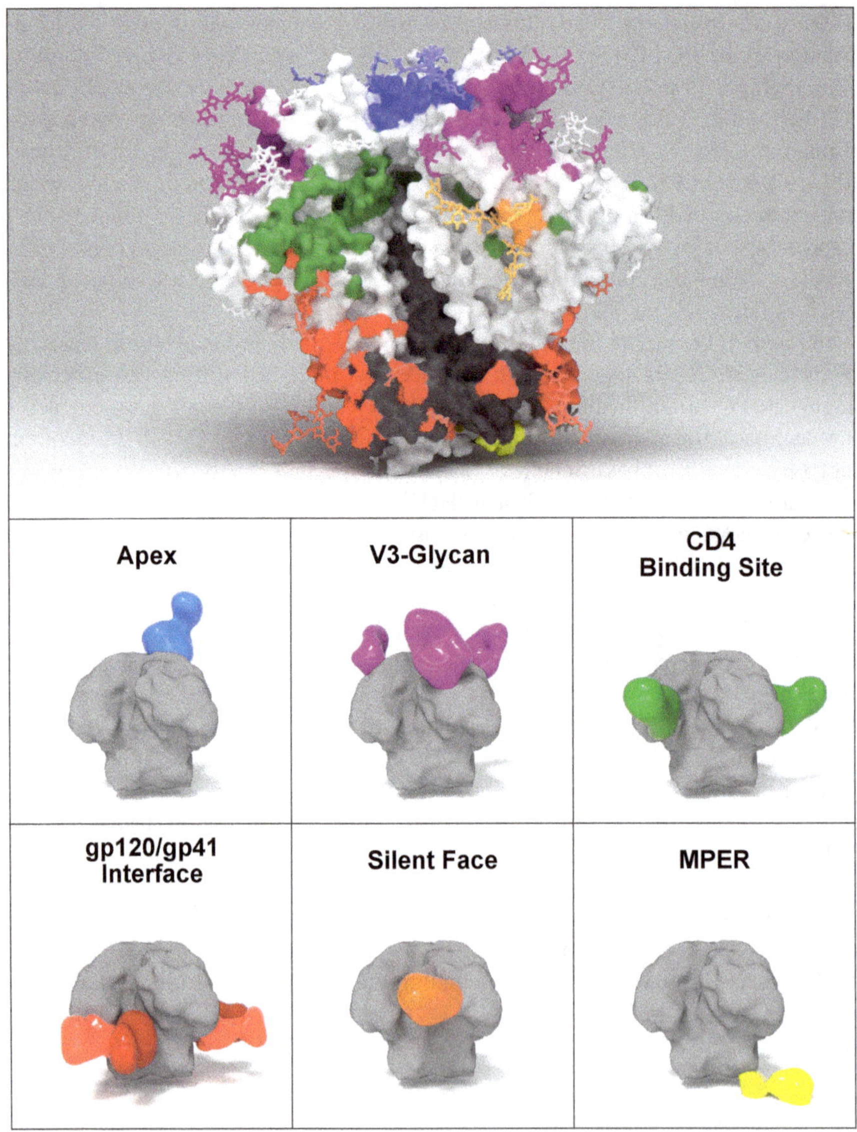

◄**Fig. 3** Recognition of HIV envelope trimers by broadly neutralizing antibodies. Top panel: Surface regions and glycans recognized by broadly neutralizing antibodies to the apex (blue), the high-mannose patch (or V3-glycan; purple), the CD4 binding site (green), the silent face (orange), the gp120/gp41 interface (red) and the MPER region (yellow). gp120 is depicted in white, gp41 in gray and N-linked glycans by sticks. The Env trimer coordinates from PDB 4TVP were used. Lower panels: Stoichiometry and angles of approach for representative bnAb densities from all major classes. Most antibody classes bind with a stoichiometry of 3 Abs per trimer, with the exception of apex-specific antibodies that mostly bind with a single, and occasionally two Fabs per trimer, and interface-specific antibodies where a maximal occupancy of 2 has been observed for PGT151. The density maps of PG9 (apex), PGT121 (V3–glycan), VRC01 (CD4 binding site), 8ANC195 (interface, bright red), PGT151 (fusion peptide specific, interface, dark red) are displayed in the respective panels. The density map for VRC-PG05 (orange, silent face) has been generated by aligning the gp120/VRC-PG05 co-crystal structure (PDB 6BF4) onto the BG505 Env trimer coordinates (PDB 5I8H), followed by generation of a density map by a computational low pass filtering of the complex coordinates using the EMAN 1.9 software package. The densities for 10E8 (yellow, MPER) are partially buried since it was shown that binding of this antibody requires Env to be lifted out of the membrane (Lee et al. 2016). Unless indicated differently, the EM 3D class average density maps were kindly provided by Gabriel Ozorowski and fitted into the EMDB 5782 EM Env density map (gray)

CD4 binding site, the gp120-gp41 interface and the membrane-proximal external region (MPER) (Burton and Hangartner 2016). To these five legacy epitope clusters, recently a sixth one has been added, which is comprised of the densely glycosylated 'silent face' proximal to the CD4 binding site; it is called such due to the historical absence of neutralizing antibodies targeting this region.

The reasons for conservation for these areas are now largely clear and typically linked to vital functions of the viral life cycle. The CD4 binding site, for instance, is crucially important for target cell tethering and entry (Klatzmann et al. 1984). The MPER region is involved in the membrane fusion process, in which a three-helix bundle within gp41 is formed and inserted into the host cell membrane (Salzwedel et al. 1999; Muñoz-Barroso et al. 1999). Likewise, the gp120/gp41 interface harbors parts of the fusion machinery (Kong et al. 2016b) and additionally needs to be conserved to ensure proper cleavage of pre-fusion Env (Blattner et al. 2014; Falkowska et al. 2014). The V3-glycan region was shown to incorporate parts of the CCR5 co-receptor-binding site at the base of the gp120 V3 loop, which, given the high variability of adjacent residues and glycans, explains the strong conservation among CCR5-tropic isolates (Sok et al. 2016b). Reasons for the conservation of the V2-apex site likely include both trimer-stabilizing interactions of the V2 loops as well as dense packing of basic amino acids to slightly destabilize the apex. While this may seem paradoxical, the resulting delicate meta-stability likely enables structural rearrangements preceding co-receptor-binding and spike opening upon CD4 engagement. In the following sections, we will discuss in detail the corresponding bnAbs to all mentioned epitope clusters.

2.4.1 CD4 Binding Site Epitope Cluster

Antibodies to the CD4 binding site (CD4bs; Fig. 3 green) are commonly induced during natural infection; however, the majority of these antibodies are non-neutralizing or strain-specific (Scheid et al. 2009). Binding-based mapping assays with core gp120 or RSC3 (Wu et al. 2010) detect CD4 binding site reactivity in over 80% of donor sera, but only small fraction of these sera contain nAbs that compete with RSC3 in neutralization assays (Landais et al. 2016; Lynch et al. 2012). This is likely because many Abs evolving against this epitope cluster recognize the CD4 binding site as displayed on g120 or dysfunctional forms of the Env spike. In the context of Env trimer, the CD4bs is much more occluded because of its proximity to the adjacent protomer, which requires an angled binding approach (Lyumkis et al. 2013; Zhou et al. 2015). Additionally, variable loops 2 and 5 as well as glycans surrounding the CD4 binding loop, most notably N197, N276, N362 and N461, were shown to hinder antibody recognition of this epitope cluster (Lyumkis et al. 2013; Crooks et al. 2015; Zhou et al. 2015). This underlines the importance of using native-like trimers or rationally designed immunogens when attempting to elicit bnAbs of this class.

There are two major subgroups of bnAbs targeting the CD4 binding site, the HCDR3 loop dominated binders, with bnAb b12 as its prototype, and VH-gene restricted binders, of which VRC01 is the iconic class member.

HCDR3 loop binders, as the name suggests, make most of their binding contacts to side chain residues of the CD4bs with their CDR loops, with special emphasis on their HCDR3 loop. The HCDR3 loop is elongated to $\sim$20aa for many antibodies of this class, including b12, HJ16, VRC13 and VRC16, with CH103 being the exception with a short 13aa HCDR3 loop (Burton et al. 1994; Liao et al. 2013; Zhou et al. 2015). With the exception of VRC13, HCDR3 loop binders show VH-gene mutation levels of $\sim$15% (nt), which is only half of the very high mutation rates observed for some VH1-2-gene restricted bnAbs like VRC01. Loop binders further do not induce the CD4 bound state (CD4i) on trimeric Env, which exposes a set of epitopes involving the gp120 bridging sheet that is recognized by strain-specific and non-neutralizing antibodies (nnAbs) such as 17b. On the contrary, b12, for example, binds to the relaxed trimer conformation and prevents further exposure of CD4i epitopes (Scheid et al. 2011; Falkowska et al. 2012). VRC13 is currently the most broad and potent member of this class (Zhou et al. 2015).

VH-gene-restricted CD4bs bnAbs, on the other hand, have shorter HCDR3s and engage with the CD4 binding site in a sCD4-like fashion, largely using residues of the HCDR2 loop and HC framework regions to form hydrogen bonds with main-chain atoms of the CD4 binding site (Zhou et al. 2010, 2013, 2015). VH-gene-restricted bnAbs can be sub-divided into VH1-2-derived antibodies, featuring VRC01, 12A12, NIH-45-46, PGV04, 3BNC117 and the more recently isolated N6 (Wu et al. 2010; Scheid et al. 2011; Falkowska et al. 2012; Huang et al. 2016) and bnAbs utilizing the related VH1-46 gene, including 8ANC131, 1B2530, as well as the CH235 lineage (Scheid et al. 2011).

Aside from very high heavy chain SHM levels needed to enable CD4-like binding, VH-gene-restricted bnAbs require modifications to their CDRL1 binding loops to prevent clashes with the conserved N276 glycan, which is part of the 'glycan fence' encircling the CD4 binding site (Zhou et al. 2013). Both VH1-2 and VH1-46 class bnAbs solve this problem by either deleting 3-6 CDRL1 residues, or by mutating 1-3 glycine amino acids into their CDRL1 loops, thus increasing flexibility (Zhou et al. 2013). The only notable exception is CH235.12, the most potent member of the CH235 lineage (Bonsignori et al. 2016), which has a naturally short 6aa CDRL1 derived from the VK3-15 gene. The binding angle of VH1-2-restricted bnAbs further requires unnaturally short CDRL3 loops that are 5aa in length to accommodate and bind loop D residues (Zhou et al. 2013). VH1-46-restricted bnAbs, on the other hand, approach the CD4 binding site slightly rotated compared to VH1-2 group antibodies, thus avoiding CDRL3 length constraints (Zhou et al. 2015). The more recently isolated IOMA bnAb presents an interesting hybrid between both classes; although it is VH1-2 derived, a unique interaction between residue D93 in its 8aa CDRL3 and gp120 N280 (loop D) allows IOMA to wedge its long CDRL3 between the gp120 loop D and V5 loop, mildly displacing the V5 loop in the process (Gristick et al. 2016).

To develop some of the mentioned rare genetic features through serendipitous germline gene re-arrangement and somatic hypermutation (SHM) may take a long time during natural infection (Kong et al. 2016a), which is mirrored by the elevated share of CD4 binding site targeting broadly neutralizing donors in cohorts with longer median infection periods (Simek et al. 2009; Walker et al. 2010). However, VH-restricted CD4bs bnAbs can reach up to 98% breadth on global isolates, with potencies that are on average an order of magnitude higher than HCDR3 loop binders (with the exception of the highly mutated VRC13, which neutralizes 82% of global isolates at 0.09 µg/ml median IC_{50}) (Zhou et al. 2015). Likewise, VH1-2-derived bnAbs are typically broader and significantly more potent than VH1-46-derived bnAbs, the notable exception here being CH256.12, which neutralizes 90% of global isolates at 0.6 µg/ml median IC_{50} (Zhou et al. 2015). Nevertheless, the undisputed new paragon of CD4bs bnAbs, succeeding VRC01, is the VH1-2-derived N6 antibody, which neutralizes 98% of global isolates at the very high potency of 0.04 µg/ml median IC_{50} (Huang et al. 2016). The high neutralization breadth and decent potency of VH1-2-restricted CD4bs bnAbs makes this class of antibodies an attractive target for reverse vaccine design strategies; however, their rare genetic features require rational design approaches to accelerate maturation, which will be discussed in Sect. 4.

2.4.2 Membrane-Proximal External Region (MPER) Epitope Cluster

MPER-directed bnAbs are observed rather infrequently in natural infection (Walker et al. 2010; Landais et al. 2016), although this epitope was found to be targeted in 27–40% of sera by strain-specific nAbs (Huang et al. 2012a; Landais et al. 2016). In a recent analysis of the Protocol C cohort, up to 40% of donors were able to neutralize HIV-2 chimeric viruses harboring the HIV-1 MPER epitope cluster; however, only 10% of MPER-targeting sera, or 2.5% cohort-wide, developed neutralization breadth (Landais et al. 2016), which is in agreement with previous reports (Binley et al. 2008; Walker et al. 2010; Gray et al. 2011). Reasons for the frequent development of strain-specific nAbs to this epitope cluster in natural infection may relate to the overall high frequency with which the adaptive immune response targets gp41 early in infection and following vaccination (Tomaras et al. 2008). It was proposed that preexisting microbiota-targeting B cell clones facilitate fast development of typically non-neutralizing antibody responses to gp41 in humans and rhesus macaques (Williams et al. 2015; Han et al. 2017), while in the process, however, outcompeting rarer, potentially protective B cell clones.

All MPER bnAbs target overlapping areas on an amphiphatic α-helical peptide sequence at the base of gp41, adjacent to the plasma membrane. This region is involved in regulating the assembly of the fusion protein and thus highly conserved, which explains the extraordinary neutralization breadth of up 98–99% that MPER bnAbs 4E10, 10E8 and DH511.2 display (Huang et al. 2012a). However, the first generation of MPER bnAbs, 4E10, 2F5 and Z13e1 (Buchacher et al. 1994; Zwick et al. 2001), showed fairly limited potency, which raised the question whether high enough serum titers to achieve sterilizing protection could be induced by vaccination. Notably, both 2F5 and 4E10 protected from infection in passive transfer studies against the neutralization sensitive SHIV Ba-L (Hessell et al. 2010). The isolation of the over fivefold more potent (0.35 µg/ml median IC_{50}) 10E8 in 2012, reinvigorated efforts to design MPER-based vaccines (Huang et al. 2012a). The recently isolated DH511.2 shows intermediate potency of 1.0 µg/ml median IC_{50}, thus falling between 10E8 and 4E10 (Williams et al. 2017).

In order to efficiently bind to their epitope, MPER bnAbs use elongated HCDR3s (20–24aa) that are rich in tryptophan and other hydrophobic residues, which allows for hydrophobic packing with both the peptide and the plasma membrane itself. MPER bnAbs were in fact long thought to be able to bind to membrane phospholipids (Alam et al. 2009) in addition to the MPER peptide, which was shown for both 4E10 and 10e8 by x-ray crystallography (Irimia et al. 2016, 2017).

The long hydrophobic HCDR3 loops that enable lipid reactivity, however, also facilitate binding to ubiquitous phospholipids, which renders this class of bnAbs prone to develop poly- or autoreactivity. Indeed, antibodies 4e10 and many members of the DH511 lineage were shown to be polyreactive (Haynes et al. 2005; Mouquet et al. 2010; Liu et al. 2014), and mouse knock-in models of bnAbs 4E10 and 2F5 showed strong depletion of mature B cells, due to negative selection of autoreactive B cell clones in the bone marrow (Doyle-Cooper et al. 2013) (Verkoczy et al. 2010, 2011; Doyle-Cooper et al. 2013). Importantly, bnAbs 10e8 and DH511.2 show no signs of poly- or autoreactivity (Huang et al. 2012a; Liu et al. 2014), which taken together with the increased potency of second-generation MPER bnAbs has made this bnAb class a target for both rational immunogen design and passive immunotherapy (Irimia et al. 2017).

A challenge for the induction of MPER antibodies is that the faithful reproduction of the epitope might require the presence of proximal lipids on the immunogen (Irimia et al. 2016), which may require liposomal delivery or related strategies liked lipid disks. Lastly, MPER bnAbs frequently show incomplete neutralization of HIV isolates (Kim et al. 2014; McCoy et al. 2015), which is likely due to glycan heterogeneity of the four large complex-type glycans present on gp41 (Cao et al. 2017, 2018b) that MPER bnAbs have to accommodate. The importance of incomplete neutralization for in vivo protection is an area of active investigation, but a recent study with 3BNC117 suggests that strong neutralization plateaus of $\sim 70\%$ maximum neutralization percent reached against the challenge virus (using the TZM-bl assay), lessen the protective capacity of passively transferred bnAbs (Julg et al. 2017).

2.4.3 gp120-Gp41 Interface/Fusion Peptide Epitope Cluster

Over the past 4 years, an entirely new family of bnAbs has been described that targets the interface between the gp120 and gp41 subunits (Fig. 3, red). In the Protocol C cohort, $\sim 12–20\%$ of broadly neutralizing sera was shown to target interface or unidentified quaternary epitopes (Landais et al. 2016). Interface targeting bnAbs are unique in that they bind glycans and peptide residues that are spread between gp120 and gp41 and in some instances also between adjacent protomers. Unlike most bnAbs to the CD4 binding site and the trimer apex, bnAbs of this class do not necessarily compete with each other, which is a function of the wide surface area that this site of vulnerability encompasses on the trimer (Burton and Hangartner 2016).

One of the first described members of this class of bnAbs was PGT151, which binds to a complex quaternary epitope comprised of the N611 and N637 glycans of two adjacent gp41 protomers, as well as residues 499-503 on gp120 adjacent to the furin cleavage site (Blattner et al. 2014; Falkowska et al. 2014). PGT151 is thus able to discern between cleaved and uncleaved forms of Env, a unique feature that has since been harnessed to assess cleavage efficiency for recombinant Env trimers. Upon binding of fully cleaved Env, PGT151 stabilizes the trimer in its native, closed conformation, which was recently leveraged to extract and structurally characterize native, full-length Env trimers, only mildly truncated in their C-terminal domain, without further stabilization (Lee et al. 2016). Both PGT151 and another gp41-gp-41 inter-protomer binding bnAb, 3BC315, induce structural rearrangements upon binding that reduced the maximal occupancy to two Fabs bound per trimer (Blattner et al. 2014; Lee et al. 2015). All other described bnAbs of this class can engage the trimer in a 3:1 stoichiometry (Huang et al. 2014; Scharf et al. 2015; Kong et al. 2016b; Wibmer et al. 2017).

Similarly, bNAb 35O22 also stabilizes the trimer upon binding, although its epitope is more central and membrane-proximal compared to PGT151. It binds the gp120 glycans N88, N230 and N241 as well as glycan N625 on gp41, but, unlike PGT151, is not cleavage sensitive (Huang et al. 2014). Both PGT151 and 35O22 show exceptional potency of 0.008 and 0.03 µg/ml median IC_{50} across large virus panels, with decent breadth of $\sim$65%.

A third member of this class, 8ANC195, originally isolated using a gp120 probe (Scheid et al. 2011), was shown to bind to glycans N234, N276 on gp120 and also glycan N637 on gp41 (Scharf et al. 2014), although the latter is not strictly required. It has the ability to bind both open, CD4 bound and closed states of pre-fusion Env, but induces a more closed conformation upon binding, even in the presence of CD4, which likely prevents full opening and fusion of the trimer (Scharf et al. 2015).

Another set of previously isolated bnAbs, 3BNC315 and 3BNC176 (Klein et al. 2012), was shown to target an epitope partially overlapping that of 35O22; however, they were shifted even further toward the base of the trimer (Lee et al. 2015). In contrast to PGT151/35O22, these antibodies destabilize the trimer and accelerate decay as a mechanism of neutralization, which is unique among bnAbs targeting this epitope cluster. They are also unique in that they are the only bnAbs of this class that do not strictly require glycans to bind (Lee et al. 2015).

The latest additions to this evolving family of bnAbs are ACS202 and VRC34, two bnAbs that directly bind the fusion peptide at the N-terminal end of gp41 (aa512–525) as well glycan N88 on gp120 (Kong et al. 2016b). Early comparisons indicate that these bnAbs target remarkably similar epitopes, which are located in between the PGT151 and 35O22/3BNC315 binding sites. Likewise, their neutralization breadth and potency is almost identical at $\sim$0.2 µg/ml and $\sim$45% breadth, which is more potent than 8ANC195 0.81 µg/ml (Kong et al. 2016b), but less potent than PGT151 and 35O22. For VRC34.01, major escape mutations are located in the fusion peptide. Both ACS202 and VRC34, similar to other bnAbs targeting the N-terminal end of gp41, are cleavage-sensitive but, unlike PGT151 and 3BC115, can bind the trimer in a 3:1 stoichiometry (Kong et al. 2016b).

All glycan-binding bnAbs of this class, including the fusion peptide binders, show high incidence of incomplete neutralization on diverse global isolates (McCoy et al. 2015; Wibmer et al. 2017), which is, as for MPER bnAbs, linked to the complex and sometimes incomplete glycosylation of gp41 (Cao et al. 2017, 2018b). PGT151, for example, was shown to bind tri- and tetra-antennary glycans at position N611/N637 (Falkowska et al. 2014) and had the highest proportion of viruses neutralized with less than 80% maximal neutralization among a diverse panel of bnAbs (McCoy et al. 2015). The worst case of a broadly but severely incompletely neutralizing antibody is the recently isolated CAP248-2B antibody, which often times fails to reach 50% maximal neutralization on large virus panels (Wibmer et al. 2017). Its expansive epitope stretches all the way from the plasma membrane/MPER region to the gp120-gp41 interface and is thus prone to be affected by changes in N-glycan glycoforms and variable residues included in its binding footprint.

2.4.4   V3-Glycan Epitope Cluster

The V3-glycan epitope cluster is centered around the N-linked glycan at position N332, which is positioned at the base of the gp120 V3-loop and surrounded by multiple oligomannose-type glycans that collectively form the high-mannose patch (Doores et al. 2010; Pritchard et al. 2015a, b). The core epitope of V3-glycan bnAbs (Fig. 3, purple) is comprised of both glycans, most notably N332 itself, as well as conserved protein residues at the base of the V3-loop which comprises parts of the CCR5 co-receptor-binding site, thereby explaining both the dense glycan shielding and strong conservation among global isolates (Sok et al. 2016b). N332 glycan-dependent broadly neutralizing serum specificities are the most frequently observed serum specificity throughout several cohorts, ranging from 25 to 40% of donors (Walker et al. 2010; Gray et al. 2011; Landais et al. 2016).

The first glycan-dependent bnAb isolated was 2G12 (Trkola et al. 1996), which was shown by mutagenesis to bind strongly to the N295, N332, N339 and N392 glycans (Scanlan et al. 2002). 2G12 is unique in that it is the only known bnAb that exclusively binds to high-mannose glycans (Wang et al. 2008) without making protein contacts. As discussed in the introduction this is no small feat, as low amounts of binding energy derived from protein–glycan interactions generally dictates that glycan-dependent bnAbs must bury large amounts of surface area to gain high affinity, even when additional protein contacts are made (Kong et al. 2013). 2G12, by virtue of a rare genetic set of mutations, shows a domain exchange between its Fab arms, thus allowing 2G12 to gain affinity from binding one large continuous glycan epitope, using both its Fab arms juxtaposed in a parallel orientation (Calarese et al. 2003; Murin et al. 2014). Nevertheless, 2G12 shows fairly weak neutralization potency and breadth (2.4 µg/ml and 32%, respectively) (Walker et al. 2011) as a consequence of its glycan-only binding mode, comprising four core and two alternate glycans, the latter of which are bound in a strain-specific manner (N386 and N448) (Walker et al. 2011; Murin et al. 2014).

When the most prominent bnAb family of the V3-glycan class, PGT121, alongside the PGT128 and PGT135 families, was isolated in 2011, the extraordinary neutralization potency of 0.02 µg/ml and high breadth of 70% of global isolates generated great interest in this epitope cluster for vaccine design (Walker et al. 2011). BnAbs to this epitope cluster typically bind to a core glycan, either N332 or N301, multiple alternate glycans, as well as the 324-GDIR-327 peptide stretch at the V3-loop base (Pejchal et al. 2011; Julien et al. 2013a; Sok et al. 2014a). Alternate glycan sites bound by V3-glycan bnAbs include N295 at the base of the V3-loop, N137 and N156 in the V1/V2-loop, as well as the N386 and N392 glycans in the V4-loop. PGT128, for example, binds the base of the N301 glycan as its core glycan and can utilize glycans N301 and N332 as alternate glycans on most HIV isolates. While the majority of glycans in this epitope cluster are of the high-mannose type, N137 and N301 are at least partially hybrid- or complex-type glycans on virions (Cao et al. 2018b). Some V3-glycan bnAbs can tolerate the removal of an alternate glycan site without loss of neutralization potency, but the removal of their respective core glycan, or the removal of multiple alternate glycans sites simultaneously, typically causes a decrease or loss of neutralization potency. This plasticity in glycan recognition has been referred to as promiscuous glycan binding (Sok et al. 2014a). Some of the most potent V3-glycan bnAbs, i.e., PGT121 and PGT128, can even tolerate loss of or shift (e.g., N332 to N334) of their core glycan, and thus also neutralize some isolates naturally lacking respective core glycans in a strain-specific manner (Sok et al. 2014a). Medium potency V3-directed bnAbs, like the recently isolated PGDM11 and PGDM21 families or extremely N332-dependent antibodies like 10-1074 and PGT124, both clonal relatives of PGT121, cannot tolerate shifts or loss of core glycans (Mouquet et al. 2012; Shingai et al. 2013; Sok et al. 2014a; Garces et al. 2014). Similarly, there are qualitative differences in the recognition of the 324-GDIR-327 peptide sequence between high- and low-potency V3-glycan bnAbs. For instance, PGT121 makes multiple side-chain contacts with residues D325 and R327 and is not negatively impacted by loss of either single site, while PGDM21 and 11 make contacts with either D325 or R327, respectively, and are thus more prone to viral escape by single amino acid substitutions (Sok et al. 2016b). In summary, V3-glycan bnAbs with higher neutralization potency and breadth have more redundancy in epitope recognition, which is the molecular-level explanation for their ability to broadly neutralize a somewhat variable epitope cluster. Both allosteric inhibition of CD4 binding and accelerated decay of infectious viral particles has been discussed as the mechanism of neutralization for this class of bnAbs (Pejchal et al. 2011; Julien et al. 2013b).

Unlike CD4 binding site directed bnAbs, V3-glycans bnAbs are generated from a heterogeneous set of VH genes, which is mirrored by diverse angles of approach that have been described for this bnAb class (Julien et al. 2013a). The malleability in epitope recognition of the high-mannose patch is likely related to the high incidence of N332-dependent donor sera in long-term HIV infected cohorts and thus attractive to vaccine design (Walker et al. 2010; Landais et al. 2016). Nevertheless, many V3-glycan bnAbs often bear insertions and deletions (indels) which occasionally arise during affinity maturation in germinal centers (Sok et al. 2013; Kepler et al. 2014). Like V2-apex targeting bnAbs, V3-glycan bnAbs utilize long HCDR3 loops, varying from 20 to 26aa in length, to penetrate the glycan shield and reach the underlying peptide surface (Burton and Hangartner 2016).

Two recently isolated V3-glycan bnAbs, BG18 and DH270.6, both show strong median neutralization potencies (0.03 µg/ml and 0.08 µg/ml, respectively) and breadth (64% and 55%, respectively) on large virus panels, that is on par with PGT121 and PGT128 (Freund et al. 2017; Bonsignori et al. 2017a). Unlike the PGT families, however, neither BG18 nor DH270.6 bear indels and both utilize shorter HCDR3 loops that are 21 and 22 aa in length, respectively. DH270.6 displays comparatively low levels of somatic hypermutation in its VH gene (~12% of nucleotides), reaffirming the suitability of the high-mannose patch epitope cluster for vaccine design. Two studies further detailed the development of V3-glycan bnAbs over multiple years of infection, thus enabling lineage-based immunization approaches to this site of vulnerability on Env (MacLeod et al. 2016; Bonsignori et al. 2017a).

2.4.5 V2-Apex Glycan Epitope Cluster

The trimer apex comprises another quaternary epitope cluster that is conserved among global isolates. However, unlike the gp120/gp41 interface, the core epitope is well defined and comprises glycans of the gp120 V2 loop as well as a stretch of basic amino acids on gp120 strand C (McLellan et al. 2011; Andrabi et al. 2015) at the apex of the trimer. Thus, bnAbs of this class are collectively referred to as V2-apex targeting (Fig. 3, blue). V2-apex bnAbs have been reported in 10–20% of long-term HIV-infected donors throughout multiple cohorts (Walker et al. 2010; Landais et al. 2016).

V2-apex bnAbs strongly overlap in their binding footprint, as they typically bind the trimer close to or along the C3-symetry axis in a 1:1 stoichiometry, but important differences in their exact binding approach can be discerned (McLellan et al. 2011; Julien et al. 2013c; Pancera et al. 2013; Andrabi et al. 2015; Gorman et al. 2016). bnAbs of this class target a relatively conserved epitope that is comprised of the N156 and N160 glycans on the gp120 V2 loop, and a stretch of basic, lysine-rich amino acids, 168-KKQK-171, on gp120 strand C (McLellan et al. 2011; Andrabi et al. 2015; Gorman et al. 2016). The bnAbs bind to this epitope by inserting their HCDR3 loop in between two N160s of adjacent protomers and/or nearby N156/173 glycans of adjacent protomers to form beta-sheet interactions with strand C residues, typically forming numerous hydrogen bonds. In addition, anionic amino acids in the HCDR3 loop, and in some cases 1–2 sulfated tyrosines, form salt bridges with the underlying cationic strand C lysines generating a large amount of binding energy. To penetrate the glycan shield and contact the underlying strand C residues, V2/apex bnAbs rely on the longest HCDR3 loops among all bnAb classes, ranging from 30 to 39aa in length. As is expected, most V2/apex bnAbs are highly selective for trimeric Env, although some bnAbs such as PG9 and VRC38 can also bind to gp120 of select isolates (McLellan et al. 2011; Cale et al. 2017).

The first members of this class, PG9 and PG16, were shown to asymmetrically interact with apex, having a slightly bend angle of approach from the trimer axes (Walker et al. 2009; Julien et al. 2013c). As more antibodies of this class were discovered, most notably the PGT145, CH01 and CAP256-VRC26 families, various angles of approach have been described, with some bnAbs binding straight along the trimer axes, e.g., PGT145 (Walker et al. 2011; Doria-Rose et al. 2014). All members of this class, except CH01, show very high potency ranging from $\sim$0.01 to $\sim$0.06 µg/ml median IC$_{50}$'s. Two more recently isolated bnAbs from the PGT145 and CAP256 donors, PGDM1400 and CAP256-VRC26.25, respectively, have developed even higher potencies of 0.005 µg/ml (83% breadth) and 0.001 µg/ml (69% breadth) median IC$_{50}$ values, thus representing the most potent bnAbs isolated from human donors (Sok et al. 2014b; Doria-Rose et al. 2015).

Interestingly, the more recently isolated VRC38 and BG1 bnAbs (Cale et al. 2017) both developed novel binding approaches to the V2-apex epitope cluster that defy many of the above 'rules of engagement.' Both VRC38 and BG1 bind slightly off-center beside the trimer axis with little inter-protomer engagement (Cale et al. 2017). This allows both bnAbs to bind the trimer with a 2:1 stoichiometry, which was observed by electron microscopy (EM); however, unlike BG1, VRC38 shows an up 170-fold increase in neutralization potency when Fab and IgG neutralization IC_{50} values were compared (Cale et al. 2017). Taken together with EM observations of IgG complexed Env trimers, these findings strongly suggest that VRC38 IgG can cross-link adjacent protomers of the same trimer using its two Fab arms, thus gaining in avidity. Furthermore, as these bnAbs do not have to reach past N160 glycans of adjacent protomers in the apex center, they manage to bind strand C residues by inserting considerably shorter 16-22aa HCDR3 loops between the N156 and N160 glycans of a single gp120 subunit. The novel binding approach limits the core epitope to the N-terminal strand C residues, allowing both VRC38 and BG1 to tolerate anionic residues at positions 168 and 169, respectively, which otherwise completely abrogate neutralization of traditional V2-apex bnAbs (Andrabi et al. 2015; Doria-Rose et al. 2015). Nevertheless, both VRC38 and BG1 show comparably low neutralization potency and breadth (0.54 µg/ml and 0.67 µg/ml with 30% and 37% breadth, respectively), likely due to weaker strand C interactions and increased exposure to variable residues (Cale et al. 2017).

The exceptional neutralization potency and breadth of trimer axis-centered V2-apex bnAbs, paired with comparatively moderate VH mutation frequencies of $\sim$12–16% displayed by many bnAbs of this class, renders the trimer apex especially relevant for vaccine design (Andrabi et al. 2015). At the same time, extraordinarily long HCDR3s are required for broad and potent neutralization of this epitope cluster. HCDR3 length is primarily regulated during germline recombination of V-D-J gene segments, with only certain D-J gene segment combinations enabling HCDR3 loops of 30+ aa length, thus limiting the pool of suitable germline pre-cursor B cells targeting the epitope cluster.

2.4.6 Silent Face Epitope Cluster

This epitope cluster is located on the gp120 outer domain, one of the most densely glycosylated areas on the Env trimer (Fig. 3; orange). As such, this area was for the longest time deemed immunologically silent due to a paucity of neutralizing antibodies targeting the region and was therefore dubbed 'silent face' (Wyatt et al. 1998). This only recently changed with the isolation of VRC-PG05 using RSC3 sorting-probes from IAVI donor #74 (Zhou et al. 2018); the same donor from whom the CD4bs bNAb VRC-PG04 antibody had previously been isolated (Falkowska et al. 2012; Zhou et al. 2018).

The VRC-PG05 epitope consists of two main high-mannose glycans, N262 and N488, which comprise about 35% of its binding footprint, respectively, as well as glycan-covered protein residues, specifically E293, and peripheral contacts to the N295 glycan (Zhou et al. 2018). Remarkably, unlike other glycan shield penetrating antibodies of the V3 or apex epitope clusters, this bNAb matured using only mildly mutated heavy and light chain genes (9% and 6% of nucleotides, respectively), and is using an average length HCDR3 loop of 17 amino acids, which raised hopes that this epitope cluster may lend itself to less sophisticated vaccine design approaches.

Unfortunately, the heavy reliance on glycan binding, second only to 2G12, renders VRC-PG05 both lacking in potency and susceptible to escape (which we discuss in detail in Sect. 2.2), which were determined to be 0.8 µg/ml and 27% on a 208-virus panel, respectively.

Interestingly, however, Schoofs and colleagues isolated a new family of somatically related bNAbs to the silent face, which is characterized by its most potent member, SF12 (Schoofs et al. 2019). SF12's neutralization breadth and potency are somewhat higher at 0.2 µg/ml and 62% on a 119-virus panel, respectively; its heavy and light chain are also considerably more mutated however ($\sim$17% of nucleotides) and its HCDR3 loop is 23 amino acids long. Nevertheless, SF12 binds largely the same epitope as VRC-PG05, albeit from a different angle, and is heavily reliant on the presence of glycans N262 and N488 while likewise accommodating glycan N295. Its longer HCDR3 loop, as well as HCDR 1 and 2 loops, forms extensive and distinct binding contacts with the protein surface, which comprises 25% of SF12's binding footprint, as compared to merely 12% of VRC-PG05's—which likely explains the difference in neutralization breadth and potency. Due to the differences in heavy chain contacts, SF12 does not bind E293, but forms ionic bonds to R444 instead, absence of which does halve its neutralization potency. Both VRC-PG05 and SF12 show remarkable neutralization breath against Clade AE viruses, which could make them a valuable addition to therapeutic bNAb cocktails containing V3-glycan bNAbs, which often struggle to neutralize AE viruses due to their shifted N334 glycan site. Indeed, an antibody cocktail containing SF12 was able to control viremia and experimentally infected mice (Schoofs et al. 2019); however mono-therapy is easily escaped from by abrogating or shifting the N448 glycan site (Schoofs et al. 2019), which analogously was observed in donor 74 from which VRC-PG05 was isolated (Zhou et al. 2018). VRC-PG05's mechanism of neutralization is likely interference with conformational changes in the Env trimer that are needed to bind three CD4 subunits, preceding fusion spike formation (Zhou et al. 2018); a similar mechanism is likely for SF12.

3 Heterosubtypic Antibodies to Influenza Virus

3.1 Virion Structure

Electron microscopy and cryoelectron tomography have found that influenza A virions can be divided into the five classes, depending on their internal and external composition. Besides predominantly spherical virions with a diameter of about 120 nm (ranging from 84 to 170 nm), elongated virions can be found as well, which tend to be about 20 nm thinner than the average spherical particle (Calder et al. 2010; Cusack:1985hq; Harris et al. 2006). There are three viral surface proteins present on the virion surface: Hemagglutinin (HA), which is responsible for viral attachment and entry into the host cell, neuraminidase (NA), which is a receptor-destroying enzyme that has been implicated in the release of progeny viruses, and the extracellular domain of the matrix 2 protein (Zebedee and Lamb 1988) (M2), which is a proton channel responsible for the acidification of the virion interior upon uptake into endosomes. HA and neuraminidase NA spikes are densely arranged on the virion surface, with an average center-to-center spacing of 11 nm. Influenza A virions have been quantified to contain around 300 HA and 38–50 NA spikes (Cusack et al. 1985; Harris et al. 2006), with the latter being found to tend to cluster in patches (Harris et al. 2006). As illustrated in Fig. 1, the surface of IAV is therefore considerably more densely populated with spike proteins than HIV. This dense population might have an impact on accessibility of the membrane-proximal parts of the viral surface proteins, while on HIV virions the Env spikes are so sparse that they should be well accessible from all directions.

Influenza hemagglutinin is the major surface glycoprotein on the virion of influenza viruses and is responsible for virus attachment and entry into the host cell. It is a trimer of dimers consisting of three HA1 subunits forming the globular head that harbors the receptor-binding site and three HA2 subunits forming a coiled-coil stalk-like structure that contains the fusion machinery. The HA protein is expressed as a precursor HA0 protein that is posttranslationally cleft into the N-terminal HA1 and C-terminal HA2 subunits. This proteolytical processing is required to render the virions infectious and has a crucial impact on the pathogenicity of avian influenza viruses. Highly pathogenic avian influenza (HPAI) viruses possess a multibasic cleavage site that can be cut by ubiquitous furin-like proteases. Low pathogenic avian influenza (LPAI) viruses, in contrast, exhibit cleavage sites that rely on trypsin-like proteases for processing. As these are primarily present in the intestinal tract, LPAI viruses primarily cause intestinal infections, while HPAI viruses can spread systematically in birds. Human influenza viruses have LPAI-like cleavage sites depending on trypsin-like proteases that are primarily found in the respiratory tract, such as tryptase Clara which is expressed by Clara cells in bronchiolar epithelium (Kido et al. 1999), or membrane bound proteases, such as transmembrane protease serine S1 member (2TMPRSS2) and human airway trypsin-like protease (HAT) (Böttcher et al. 2006).

Hemagglutinin trimers are responsible for viral attachment to host cells by binding to sialic acid moieties on the host cell surface. While human isolates appear to prefer sialic acids linked in a α-2,6 orientation, avian isolates prefer α-2,3-linked sialic acids, and this linkage preference has been reasoned to be the primary cause for the species preference of different IAV strains (Skehel and Wiley 2000). The receptor-binding site is located on the globular head and consists of a shallow pocket that is formed by the 130-loop, the 150-loop, the 190-helix and the 220-loop of HA1. In contrast to most other viral attachment protein/receptor interactions, hemagglutinin only has a very low affinity for its sialic acid receptor. While HIV gp120 binds CD4 with affinities in the range of 1–10 nM (Ugolini et al. 1999), the binding affinity of hemagglutinin for sialic acid has been determined to be in the 2–4 mM range (Sauter et al. 1989). It is therefore presumed that numerous spike/receptor interactions are required for virus attachment. Also for the formation of the fusion pore, the most recent model found that at least six hemagglutinin molecules are required, three of which have to undergo a conformational change (Dobay et al. 2011). Following attachment to cells, IAV is internalized into endosomes, and fusion of the virus with the endosomal membrane is induced by the acidification of the endosome (Bullough et al. 1994; Skehel et al. 1982).

Both, the NA and the HA proteins exist in various subtypes that classically have been defined as serotypes. There are currently 18 subtypes described for HA, and 11 for NA. Genetically, subtypes differ in more than 25% of their amino acid sequence from each other but can diverge by more than 50% in their nucleic acid sequence. IAV can infect a broad variety of hosts but aquatic birds are considered as the main reservoir. While avian IAV strains can express most subtypes (H1–H16 and N1–N9), human viruses are restricted to the H1, H2, H3, N1 and N2 subtypes. The recently discovered H17, H18, N10 and N11 subtypes have exclusively been identified in influenza-like viruses isolated from bats (Tong et al. 2012, 2013) and use MHC-II as a cross-species receptor (Karakus et al. 2019). Equine, canine and piniped IAV viruses also only display a limited number of serotypes (Parrish et al. 2015), which makes aquatic birds the genetically most diverse reservoir of influenza A viruses (Fig. 4).

The nomenclature of all influenza viruses starts with the genus, the host species, the site of isolation, the isolate number and the year of isolation, all separated by slashes. The subtype composition is added in brackets to the end of these strings. For human viruses, the host species designation is omitted. Accordingly, A/Chicken/Germany/N'/1949 (H10N7) refers to the virus N' that was isolated from a

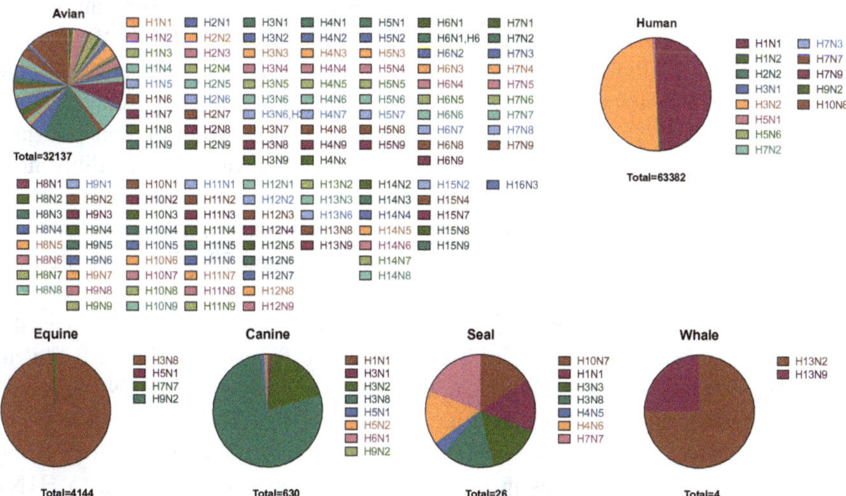

Fig. 4 Subtype distribution in different species. The NCBI Flu database was queried on September 21, 2017, for unique HA and NA sequences from the indicated species and the number of summed up HA and NA sequences were plotted for each indicated host species

chicken in Germany in the year 1949, and that expresses an HA of the H10 and a NA of the N7 subtype.

The segmented genome of IAV resides on eight distinct, negative-sensed RNA molecules. During superinfection of cells, these molecules can be exchanged for each other giving rise to new, reassorted variants. IAV have also a limited ability to be transmitted across species barriers, allowing for the introduction of new subtypes from one species to another. In the case of humans, zoonotic infections from poultry are usually abortive, largely benign and only cause minor symptoms such as conjunctivitis. However, since the late 90s, a rising number of zoonotic infections with HPAI viruses of the H5N1 subtype have been observed. Although these viruses can typically not be transmitted from human to human, their very high lethality rate of over 50% is a major concern (http://www.who.int/influenza/human_animal_interface/H5N1_cumulative_table_archives/en/). Also, infections with HPAI viruses of the H7N9 subtype have become more frequent in recent years.

Successful introduction of a new subtype into the human IAV requires species transgression that is follow by the adaption of the transgressed virus to the human host. For instance, a reassorted virus with an avian HA gene has to gain the ability to use the α-2,6-linked sialic acid as a receptor, needs to balance with the remaining genes such as the receptor-destroying neuraminidase gene (Scholtissek et al. 2002; Wagner et al. 2002; Ilyushina et al. 2005) and has to acquire the ability to efficiently transmit between humans. Due to their novel antigenic composition, such reassortant viruses constitute an antigenic shift that typically leads to an influenza pandemic. Since swine express both α-2′,6′ and α-2′,3′-linked glycans in their lungs, they can be infected by both avian and human viruses, and are considered to be the mayor mixing vessel in which avian viruses reassort with human viruses, and gain the ability to use α-2′,6′-linked sialic acid receptors. These viruses can then be transmitted to humans and cause a new pandemic. However, the H1N1 'Swine flu' pandemic indicated that such reassortment is not absolutely required and a porcine virus can directly adapt to the human host. There is considerable concern that zoonotic avian H5N1 HPAI eventually could achieve a similar adaption and gain the ability for efficient human-to-human transmission. Given the incredibly high cumulative lethality rate of over 50% of infected individuals, the effects of such an adaption would be devastating. For comparison, the 'Spanish Flu' (see below) had an estimated lethality rate of 1.5% and killed an estimated 50 million people over two years (De Jong et al. 2000).

The twentieth century has seen four major antigenic shifts. The first one, the introduction of H1N1-expressing viruses, caused the devastating 'Spanish Flu' in 1918. H1N1 viruses continued to circulate until a virus of the H2N2 serotype replaced them during the 'Asian Flu' pandemic of 1957. H2N2 viruses themselves were then replaced by a reassortant H3N2-expressing virus during the 'Hong Kong Flu' in 1967. In 1979, viruses that are genetically very closely related to H1N1 viruses circulating in the late 50s caused the 'Russian Flu' pandemic (Nakajima et al. 1978). However, these H1N1 viruses failed to replace the circulating H3N2 viruses. It is currently assumed that the 'Russian Flu' originated from a failed vaccination experiment using an insufficiently attenuated virus derived from an isolate circulating in the late 50s. Ever since the introduction of this virus in 1979, both the H1N1 and H3N2 subtypes co-circulate in parallel and evolve independently from each other. This dual subtype circulation was also not resolved during the latest (pseudo-)pandemic caused by a species-transgressed H1N1-expressing porcine virus that caused the 2009 'Swine flu.' However, the novel H1N1 viruses replaced the previously circulating H1 lineage.

3.2 Measures of Immune Evasion (Spacing, Glycosylation, Hidden Epitopes)

3.2.1 Escape Within the Population Versus Within Single Host (Mutation Rates)

In contrast to HIV, which persists within a host and incorporates measures to escape immune pressure exerted by the host immune system, IAV evolves much slower. Their primary selection pressure arises from a dwindling number of non-immune hosts that selects for variants resistant to herd immunity, i.e., the abundance of neutralizing antibodies to circulating variants in the population. This process, referred to as antigenic drift, enables IAV to infect hosts who have previously been infected by the same subtype and ensures continuous circulation of the virus. Antigenic shift, and the short-lived antibody titers, is at the base of the seasonal influenza epidemics and is also responsible for the need to annually revisit the IAV vaccine formulation. Interestingly, antigenic shift is mainly observed in circulating human viruses. In contrast, porcine viruses are antigenically considerably more stable than human IAV (De Jong et al. 2007), and virtually no antigenic drift can be observed in circulating avian IAV (Bean et al. 1992). The difference could be explained by (i) the shorter generation span of both swine and birds which ensures that there is a constant supply of immunologically naïve offspring and (ii) in the case of avian IAV, the many available serotypes make escape from herd immunity less necessary. However, antigenic drift has been observed in H5N1 IAV and is thought to be the result of prolonged circulation in live poultry (Cattoli et al. 2011; Zou et al. 2012).

3.2.2 HA Sites of Vulnerability

Head Region

Using competition experiments with monoclonal antibodies, the classical immunodominant antigenic sites of human IAV HA have been determined. For H1, they are referred to as Ca1, Ca2, Cb, Sa and Sb (Fig. 5 and (Gerhard et al. 1981), for H2 as IA, IB, IC, IIA and IIB (Tsuchiya et al. 2001), for H3 as A, B, C and D (Wiley

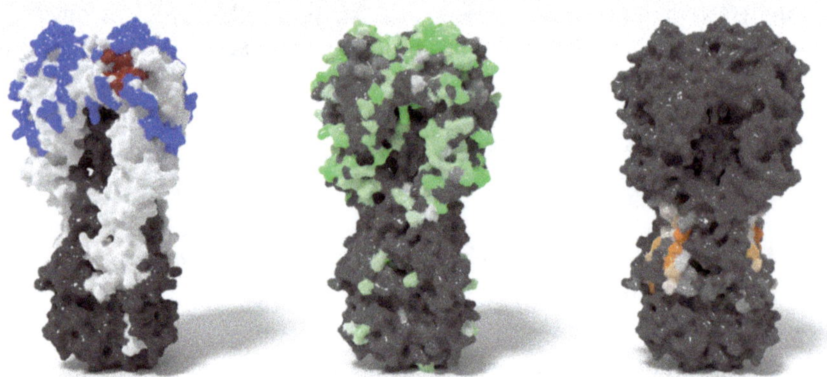

Fig. 5 HA epitope location in relation to sequence variability. Left: the classical antigenic sites, as defined by monoclonal antibodies, are depicted in blue and surround the receptor-binding site (red). Middle: Shannon's Entropy (Variably) for Human H1 between 1918 and 2001. For the variability heat-map, 8219 full-length human H1 protein sequences were downloaded from the NIH Flu database (http://www.ncbi.nlm.nih.gov/genomes/FLU/Database) and aligned using the MUSCLE algorithm. Using an aligned FASTA file, Shannon entropy was then calculated with a web-based service provided by the Los Alamos National Laboratories (http://www.hiv.lanl.gov/content/sequence/ENTROPY/entropy_one.html) and mapped on the surface structure using a custom Python script in MacPyMol. The degree of entropy is indicated by a green (highest entropy, 1.058) to gray (no entropy, 0) gradient. The maximal possible Shannon entropy for standard amino acids is ln(20) or 2.996. Right: residues recognized by stem-specific, heterosubtypic antibodies

et al. 1981), for H5 as I and II (Kaverin et al. 2007), and for H9 as I and II (Ilyushina et al. 2004), or H9-A and H9-B (Peacock et al. 2016). These antigenic sites are all located on the globular head formed by HA1, and surround or slightly overlap the receptor-binding site. In H2 (and probably H9), an additional antigenic site has been found in the stem of the protein (Tsuchiya et al. 2001). A more recent bioinformatic approach extended the definition of the antigenic sites for H1 to include more residues than the classical approach using monoclonal antibody competition (Huang et al. 2012b).

When the Shannon entropy (a mathematical measure for entropy, or unpredictability) is calculated from an alignment of over 8200 unique human H1 HA sequences, it becomes apparent that the greatest variability can be found at the classical antigenic sites, nicely illustrating that the antigenic drift is caused by the selection pressure arising from H1-specific antibodies (Fig. 5). Hemagglutinin is not as heavily glycosylated as the HIV envelope with 6 to 14 N-linked glycosylation sites per protomer. While some of these glycosylation sites are required for proper folding (Roberts et al. 1993), influenza A viruses also primarily use shifting glycosylation to reduce immunogenicity and to evade antibody recognition. Human H3N2 viruses, for instance, have almost doubled the number of N-linked glycosylation sites on hemagglutinin since their introduction into human population in

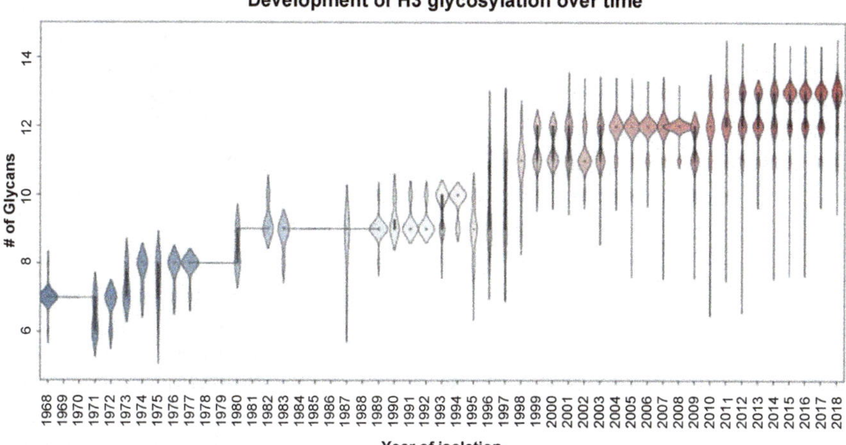

Fig. 6 Acquisition of additional N-linked glycosylation sites on the HA of H3N2 viruses. Since their introduction into humans in 1968, H3N2 viruses have almost doubled the number of N-linked glycosylation sites. Like in HIV-1, this accumulation of glycans is thought to help the virus to escape preexisting antibodies and to lower its immunogenicity. For the generation of the plot, all unique full-length human H3 sequences with known year of isolation were downloaded from https://www.ncbi.nlm.nih.gov/genomes/FLU/ (8/20/2018; 7057 sequences) and potential glycosylation sites counted with a custom Python script and plotted using the Seaborn data visualization library

1968 (Fig. 6), probably in response to pressure for preexisting antibodies in the population. However, for influenza, it has been postulated that introduction of glycans comes at a fitness cost (Das et al. 2011). Moreover, it has been simulated that there is a maximal number of supported glycans for a given configuration, and that after this number is reached, typically after 9–12 years of evolution in the human host, the glycans have to be swapped between different sites, or the hemagglutinin is replaced by a novel pandemic strain (Altman et al. 2017).

3.2.3   Broadly Neutralizing Antibodies to Influenza a Viruses

While the majority of antibodies elicited during infection are highly strain-specific, and only neutralize the eliciting, or closely related viruses, evidence has been accumulating that a small fraction of antibodies is actually capable of recognizing a wide range of viruses within one subtype, or even viruses belonging to different subtypes. The latter are therefore referred to as heterosubtypic antibodies (hnAbs). Heterosubtypic antibodies can be found in a majority of people, albeit at low titers (Li et al. 2012; Kohler et al. 2014). Heterosubtypic antibodies tend to be restricted to phylogenetic group 1 [CR6261 (Ekiert et al. 2009) F10 (Sui et al. 2009), 3.1 (Wyrzucki et al. 2014), PN-SIA49 (De Marco et al. 2012)]. Yet, hnAbs that can cope with the structural differences between the two phylogenetic groups, and Abs that are able to neutralize viruses from both groups, have been described [CR9114 (Dreyfus et al. 2012), FI6 (Corti et al. 2011), 1.12 (Wyrzucki et al. 2015), FY1/ MEDI8852 (Kallewaard et al. 2016)]. One of these hnAbs, CR9114, can also neutralize viruses of the influenza B virus genus, indicating that even differences between different IAV genera can be overcome. All these antibodies bind to a conserved hydrophobic groove that is formed between HA1 and the A helix of the HA2 subunit, and often also involve contacts to the fusion peptide (FI6, MEDI8852). Stem-specific antibodies do not interfere with receptor binding but prevent the structural rearrangements required for the fusion process (Ekiert et al. 2009). Their binding has also been described to inhibit proteolytical activation of the HA0 precursor protein (Corti et al. 2011). Within the primary stem epitopes, there are several differences, that hnAb must accommodate to neutralize viruses from both phylogenetic groups. These are First, the presence of a larger Asn instead of a Thy residue at position 49 of group 2 HA2. Second, group 1 viruses have a His at position 111 of HA2, while group 2 viruses typically have Thr or Ala at this location. Although these residues are not directly contacted, they determine the orientation of a crucial Trp residues at position 21. And third, the presence of a N-linked glycosylation site at position 38 of group 2 HA2 that constitutes another major difference between the two phylogenetic groups. Accordingly, antibodies need to be able to accommodate these differences to neutralize viruses from both phylogenetic subgroups (Ekiert et al. 2009; Corti et al. 2011; Dreyfus et al. 2012; Laursen and Wilson 2013).

Interestingly, group 1 and pan-IAV heterosubtypic antibodies primarily contact the HA stem groove using residues of the heavy chain (Ekiert et al. 2009; Corti et al. 2011). In the case of pan-IAV hnAb 1.12 that was isolated by phage display, the same heavy chain was found paired with numerous light chains and most heavy-light chain combinations resulted in hnAbs. Also of note is that the broadest pan-IAV antibodies, 1.12 and CR9114, display an unusual stretch of 4–5 tyrosine residues at the tip of their HCDR3 that is involved in antigen recognition. Indeed, antibodies capable of neutralizing both phylogenetic groups have been shown to evolve from antibodies that only neutralize viruses from phylogenetic group 1, and then acquire group 2 specificity during maturation (Kallewaard et al. 2016). Monoclonal antibodies specific only for phylogenetic group 2 have also been isolated, but they are considerably less frequent than those to group 1 [CR8020: (Friesen et al. 2014)] and bind to an epitope located at the base of the stem of the HA trimer (Corti et al. 2011; Friesen et al. 2014). Stem-specific antibodies can be compared to HIV bnAbs that bind to the interface between gp120 and gp41 such as PGT151 (Falkowska et al. 2014), 8ANC195 (Scharf et al. 2014) and 35O22 (Pancera et al. 2014), and have likewise been postulated to interfere with structural rearrangements preventing the formation of the fusion spike.

A second class of heterosubtypic antibodies has been described that mimics sialic acid receptor binding by inserting a complementarity determining region (CDR)-loop into the receptor-binding site. As for HIV, the receptor-binding site of IAV is relatively conserved in order to remain functional, and thus enables antibodies recognizing key receptor-binding residues to neutralize a diverse range of IAVs (Ekiert et al. 2012). However, in contrast to CD4 binding site-specific bnAbs to HIV, such as VRC01 or 3BNC117, these antibodies usually have a more restricted breadth, yet provide an interesting target for the design of a universal influenza vaccine.

4 Vaccine Design Aspects for HIV and Influenza

4.1 Approaches for Eliciting Broadly Neutralizing/Heterosubtypic Antibodies

It is currently assumed that there are two major obstacles for the elicitation of broadly neutralizing antibodies: First, immunodominance of the variable and strain-specific epitopes is suspected to impede antibody responses to the conserved and clearly less immunogenic h/bnAb epitopes. Second, the scarcity of h/bnAb precursor B cells whose BCR, in case of HIV, often also do not recognize the viral envelope proteins very well (Scheid et al. 2011; Ota et al. 2012; Hoot et al. 2013; Escolano et al. 2016). Moreover, studies of bnAbs to HIV show that a successful vaccine against antigenically highly variable viruses not only needs to induce antibodies that are able to recognize relatively conserved epitopes, but that are also able to tolerate variation within the epitope. This requirement is considerably more difficult to achieve than simply increasing an antibody's affinity. For instance, bnAb b12's affinity for gp120 was improved over 420-fold by in vitro affinity maturation (Barbas et al. 1994; Yang et al. 1995). Yet, despite gaining the ability to neutralize some additional viruses, the overall breadth of the affinity matured antibody decreased, indicating that breadth is a fine-tuned balance of affinity, conservation of the recognized epitope and ability to tolerate variations within the epitope.

There are currently two main immunization strategies by which these obstacles are sought to be overcome (Fig. 7): The first strategy is based on multiple immunogens that are antigenically distinct except for desired epitope(s). The hope in this approach is that the desired epitopes become a common denominator shared between all immunogen variants and will therefore provide an advantage to maturing B cell clones targeting shared over differing epitopes. Such common denominator immunizations can be performed in the form of immunogen cocktails, administering all immunogen variants at once, or in serial immunizations, by using consecutive individual immunogens.

The second strategy employs a sequence of immunogens, each of which is designed to shepherd the antibody response into a desired direction. Some of these strategies involve a priming immunization with an antigen that has been designed to activate rare h/bnAb precursor B cells, which often do not recognize the actual epitope on the pathogen very well. These 'germline targeting' immunogens are then followed by immunizations with immunogens designed to gradually shepherd the B

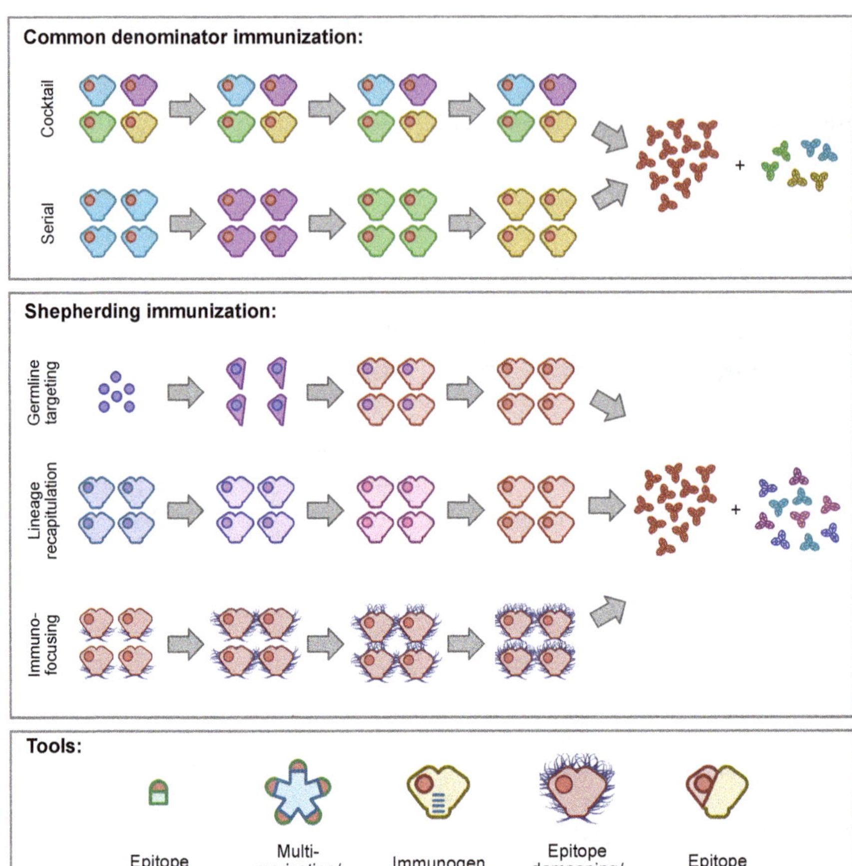

Fig. 7 Strategies to elicit broadly reactive antibodies. The 'common denominator' strategies use immunogens that are antigenically distinct yet share the desired epitope (red circle). This strategy can be pursued using immunogen cocktails, or serial immunization with different individual immunogens. The 'shepherding' strategies start with an immunogen specifically designed to activate B cells expressing the germline of h/bnAbs as BCRs, followed by booster immunizations designed to shepherd the response toward broadly reactive antibody responses recognizing the epitope, as it is present on the pathogen. Alternatively, evolution of h/bnAb lineages is recapitulated by immunizing with immunogens whose succession lead to the selection of h/bnAbs in vivo. Lastly, immunofocusing aims at first eliciting a range of antibody responses from which the desired responses are selected by sequential use of epitope-dampened immunogens, in which only the relevant epitopes are accessible to antibodies (immunofocusing). The tools employed for these approaches include epitope scaffolding, i.e., generation of a small artificial protein displaying the epitope of interest, multimerization to increase immunogenicity, stabilization of meta-stable spike proteins, epitope dampening/resurfacing by modification of the spike surface to reduce immunogenicity of undesirable regions (e.g., by the addition of glycans), and epitope grafting, where regions of interest are transferred to a related but immunologically different spike protein (e.g., the HIV V2 grafted onto SIV Env)

cell response from the artificial 'germline-recognized' epitope to the actual h/bnAb epitope present on the pathogen's surface protein. A second embodiment of the shepherding approach is mainly used for HIV and is attempting to recapitulate the viral/antibody co-evolution as it occurred in individuals who developed bnAbs. Thereby, a sequence of envelope immunogens is applied starting with the transmitter/founder virus envelope that is then followed by variants that triggered substantial shifts toward breadth in the antibody response of the infected individual. This approach requires detailed longitudinal data from h/bnAb donors. The third embodiment aims at triggering numerous antibody responses to the immunogen, but then attempts to focus the response on the desired epitopes during the subsequent immunizations, by using a sequence of gradually more and more restrictive immunogens.

4.2 Challenges and Approaches for the Induction of bnAbs to HIV

Although many vaccine trials using traditional immunization strategies attempted to elicit bnAbs through vaccination, none have been successful thus far. The reasons are many but include the choice of immunogens that did not reflect the antigenicity of the native Env trimer, most prominently soluble gp120 and non-stabilized gp140 proteins. While great progress has been made in developing soluble, trimeric Env proteins displaying native-like antigenicity in recent years (Sanders et al. 2013; Sharma et al. 2015), elicitation of bnAbs in relevant model systems is still outstanding. This is not only due to the unusual genetic features that bnAbs seemingly need to acquire over many cycles of germinal center reactions, but can also be attributed to the immunoquiescence of conserved broadly neutralized epitopes on native Env, which is rooted in the immune-evasive features of the Env spike as discussed in the bnAb section.

In short, as pointed out in the respective epitope cluster sections, bnAbs develop rare genetic features to by-pass the immune-evasive features of Env, which in turn hinders their induction by vaccination. These include (i) high levels of somatic hypermutation, which is problematic for all five major epitope clusters, (ii) insertions and deletions, most prominently among V3-glycan bnAbs, (iii) long CDR-H3 loops, particularly among V2-apex, V3-glycan and some gp120-gp41 interface targeting bnAbs, (iv) restriction to specific VH germline alleles among CD4 binding site bnAbs and (v) potentially limited pools of primary B cells able to engage epitopes that are partially autoreactive, including the glycan and lipid comprising epitopes of V2-apex, V3-glycan and especially MPER-directed bnAbs (Burton and Hangartner 2016).

Furthermore, even native infectious Env spikes on the virion surface present many immunodominant epitopes that are highly variable and thus tend to direct the immune system toward inducing strain-specific nAbs from which the virus will invariably escape. Therefore, strategies for the guided induction of bnAbs are most likely needed, either by means of engaging germline precursors of mature bnAbs, attempting to guide the immune response along a pre-defined maturation path, or by deploying other immunofocusing strategies, including sequential or cocktail immunizations with native Env trimers (Fig. 7).

4.2.1 Induction of Autologous Tier 2 nAbs by Trimeric Immunogens

The failure of clinical trials to induce tier 2 nAb responses has been associated with differences in the presentation of critical epitopes on the immunogens used, as compared to their presentation on the native Env spike. Therefore, the generation of molecules that more faithfully mimic the spike has opened up new opportunities for the induction of tier 2 nAbs. These trimeric spike mimetics include foremost SOSIP trimers (Binley et al. 2000; Sanders et al. 2002; Sanders et al. 2013), but also native flexible linker (NFL) trimers (Georgiev et al. 2015; Sharma et al. 2015) as well as the more recently described uncleaved prefusion-optimized (UFO) trimers (Kong et al. 2016a). Native-like Env trimers have successfully induced tier 2 nAbs in small animal models (Sanders et al. 2015; McCoy et al. 2016; Torrents de la Peña et al. 2017) and less reproducibly in nonhuman primates (Sanders et al. 2015).

Nonhuman primates (NHPs), and specifically rhesus macaques (RMs), are often argued as the most appropriate pre-clinical model for HIV vaccine studies because of the close genetic relatedness of NHPs to humans. Four RM studies using trimeric Env immunogens reported the induction of autologous Tier 2 nAbs, with two of those studies using SOSIP trimer designs (Sanders et al. 2015; Pauthner et al. 2017) and one using NFL trimers (Martinez-Murillo et al. 2017). However, there have been concerns about the limitations of those results. Tier 2 nAb titers were primarily reported after 6–12 months of immunizations, were relatively low, and washed out rapidly. Most worrisome was the observation that only a fraction of RMs developed nAbs, raising concerns about whether Env trimers will elicit tier 2 nAbs in humans (Havenar-Daughton et al. 2017; Sanders and Moore 2017). To address this problem, an optimized protocol for the reliable induction of tier 2 nAbs in NHPs was recently developed (Pauthner et al. 2017), which enabled the induction of tier 2 nAb titers in all immunized NHPs, and, at high titers, also protected animals from homologous SHIV challenge (Pauthner et al. 2019).

These animal studies have further uncovered that soluble spike mimetics expose highly immunogenic epitopes at the base of the trimer that are not present in the membrane inserted version of the protein (de Taeye et al. 2015; McCoy et al. 2016; Bianchi et al. 2018). Moreover, using a recently developed method to map the epitopes recognized by polyclonal antibodies by electron microscopy (Bianchi et al. 2018), it was found in NHPs that repetitive immunization with soluble proteins induced more antibody specificities than are elicited during experimental SHIV infection. Also, some of the vaccine-induced specificities were found to block access to bnAb epitopes (Nogal et al. 2020).

4.2.2 BnAb Immunization Strategies Based on Mimetics of the Native Env Spike

Because SOSIP and NFL trimers generally have excellent antigenic properties (Sanders et al. 2013; Georgiev et al. 2015; Sharma et al. 2015), meaning that nAb-binding EC_{50} and neutralization IC_{50} are closely correlated, while non-neutralizing antibodies cannot bind, there were high hopes that this class of immunogens would suffice to induce some level of neutralization breadth. However, immunizations with the first generation of native-like trimers failed to induce heterologous tier 2 nAb titers (Sanders et al. 2015; Pauthner et al. 2017). A widely discussed explanation is that native-like trimers still expose a variety of immunodominant variable epitopes that misdirect the immune system away from immunoquiescent, conserved bnAb epitopes (Havenar-Daughton et al. 2017; Pauthner et al. 2017, 2019). The immunological explanation for this phenomenon is that in germinal center (GC) reactions, B cells reactive to easy-to-engage, high-affinity epitopes expand faster than those reactive to hard-to-engage epitopes and thus outcompete the former. As a result, an immunodominance hierarchy arises, which may be further skewed by the abundance of B cells able to engage constrained, immunoquiescent epitopes, which may require rare primary or cross-reactive secondary B cells (Havenar-Daughton et al. 2017; Angeletti et al. 2017). Consequently, multiple strategies have been developed to accelerate and guide bnAb induction.

Despite the overall decent antigenicity of native-like trimers, there is evidence for some flexibility of SOSIP molecules and exposure of regions, such as the V3 loop tip, that are not or infrequently exposed on virion surface trimers (de Taeye et al. 2015). Immunizations with native soluble trimers revealed three principal targets of nnAbs that are frequently targeted: the V3 loop, epitopes exposed due to CD4-induced conformational changes (CD4i), and epitopes at the base of the trimer that are artificially exposed as a result of the truncation at amino acid 664 in the gp41 region, which renders SOSIP trimers soluble (Sanders et al. 2015). To reduce the exposure of these epitopes and thus potentially reduce competition of B cell lineages in GC-reactions, several stabilization approaches have been developed. These include designs to sequester the V3 loop (de Taeye et al. 2015; Ringe et al. 2017; Kulp et al. 2017), as well as strategies to increase the overall thermostability of the trimer to reduce the exposure of CD4i and other non-neutralizing epitopes, e.g., by adding additional disulfide bonds or by chemical cross-linking (de Taeye et al. 2015; Steichen et al. 2016; Ringe et al. 2017; Torrents de la Peña et al. 2017; Kulp et al. 2017). While these designs, as far as they have been tested in vivo, do suppress V3 loop-directed responses to various degrees, thus far none have been able to induce bnAbs or increase the titer of autologous tier 2 nAbs.

Since the biggest deterrent to efficient nAb-binding of broadly neutralizing epitopes is the density of the glycan shield, a novel immunogen design strategy is the deliberate removal of glycans shielding conserved epitopes (Crooks et al. 2017; Dubrovskaya et al. 2017) or the use of isolates with naturally occurring glycan holes (Voss et al. 2017). These immunogens were developed following two studies reporting that autologous tier 2 nAbs induced by BG505 SOSIP.664 and JRFL NFL immunizations in rabbits, were targeting naturally occurring glycan holes in the respective isolates (Crooks et al. 2015; McCoy et al. 2016). Immunogens with deletions in the glycan fence surrounding the CD4 binding site were able to induce extraordinarily high nAb titers to viruses bearing the same glycan deletions, but induction of nAbs against heterologous wild type isolates has still not been achieved (Dubrovskaya et al. 2017). This will likely require more sophisticated sequential immunization schemes that step-wise re-introduce deleted glycans and finish with wild type trimer boosts; early results using this approach look promising (Dubrovskaya et al. 2017).

Sequential and cocktail immunizations are frequently discussed immunization strategies, which rely on a heterogeneous set of immunogens that only share distinct conserved epitopes (Fig. 7). This approach was computationally predicted to give rise to germinal center dynamics supportive of the bnAb induction (Wang et al. 2015). The principal idea is that by immunizing with a trimer mix that differs in variable domains, GC B cells to conserved epitopes gain a competitive advantage over GC B cells targeting variable domains, especially when specific variable epitopes are removed completely from the immunization sequence. An easy way to achieve this is by immunizing with a heterogenous set of trimers from multiple clades, as, by definition, broadly neutralizing epitopes remain conserved while variable epitopes do not. However, there have been no examples of cross-clade trimer sequential or cocktail immunizations able to induce heterologous nAb titers thus far (Klasse et al. 2016). It is important to point out that germline-targeting approaches generally also use sequential immunization schemes, but instead of excluding variable epitopes, these strategies focus on the evolution of a specific conserved epitope.

4.2.3   bnAb Immunization Strategies Based on Germline-Targeting

Germline-targeting immunization strategies use a selection of specific immunogens to shuttle the immune response along a pre-defined maturation path. Two general approaches can be discerned: (i) so-called lineage-based approaches aim to re-create bnAb evolution as observed in the longitudinal study of infected individuals and (ii) the use of immunogens that engage specific germline versions of known bnAbs, typically by means of epitope scaffolding or rational protein design, followed by appropriate boosting regimens (Stamatatos et al. 2017) (Fig. 7).

The problem underlying both approaches is that most native trimers do not bind germline-reverted bnAb precursors (Xiao et al. 2009). However, the detailed longitudinal study of bnAb and virus co-evolution provides opportunities to identify both germline precursors of mature bnAbs as well as the viral *env* sequences that likely initiated those bnAb lineages (Bonsignori et al. 2017b). Detailed co-evolution studies have been described for bnAbs targeting the CD4 binding site (Liao et al. 2013; Gao et al. 2014), the V2-apex (Doria-Rose et al. 2014), the V3-glycan

(MacLeod et al. 2016; Bonsignori et al. 2017a) and the MPER epitope clusters (Williams et al. 2017). However, early studies using lineage-based immunogens failed to induced heterologous nAb titers in small animals and RMs (Malherbe et al. 2011; Malherbe et al. 2014). Ongoing studies using second-generation stabilized Env trimer immunogens may be more successful, but this remains to be seen.

The second germline-targeting approach is based on rationally designed immunogens that engage specific germline-precursors of mature bnAbs and has been pioneered for CD4 binding site directed bnAbs. This class of bnAbs is a convenient choice, as CD4 binding site directed bnAbs of the VRC01 class are derived from a very limited set of germline VH alleles (VH1-2 and VH1-46) and thus present a precise target for computational design. Two prototypical VRC01-class germline-targeting immunogens have emerged: 426c, developed by Stamatatos and colleagues (McGuire et al. 2013) and eOD GT6–GT8, developed by Schief and colleagues (Jardine et al. 2013). First studies using these immunogens have shown to effectively induce suitable VH1-2/5aa LCDR3 precursors in knock-in and humanized mouse models (Jardine et al. 2015; Sok et al. 2016a). However, sequential immunizations starting with a germline-targeted prime thus far failed to induce nAb titers to WT viruses with N276 glycans present. Immunogens to further guide maturation have been proposed and are underway (Briney et al. 2016). These may involve a new class of hybrid germline-targeting SOSIP trimers that show affinity to germline-precursor bnAbs in the context of a native trimer (Briney et al. 2016; Medina-Ramírez et al. 2017). Importantly, only a fraction of the mutations accumulated over years of affinity maturation is needed to effectively engage the CD4 binding site, underlining the feasibility of a rational vaccine design approach to this epitope cluster (Jardine et al. 2016b). Suitable VRC01-germline precursor B cells in naïve humans have been identified by sorting with eOD GT8 probes (Jardine et al. 2016a).

For other epitope clusters like the V2-apex and V3-glycan areas, computational design strategies have been harder to implement, partly because of incomplete epitope definition and party due to increased complexity of epitope clusters for which broad recognition involves direct glycan binding as compared to mere glycan accommodation. However, one such study using sequential immunogens designed to engage with and drive germline PGT121 maturation recently reported the induction of broadly neutralizing antibodies in germline–PGT121 knock-in mice, which represents an important proof-of-concept (Escolano et al. 2016; Steichen et al. 2016). Other recent studies repeatedly immunizing with high-mannose gly-cosylated gp140 or chemically synthesized Man$_9$-V3-peptides were able to induce V3-glycan directed antibodies in RMs (Saunders et al. 2017). However, like earlier approaches in small animals using high-mannose glycosylated yeast proteins (Zhang et al. 2015), such glycan-reactive antibodies generally fail to neutralize WT isolates that are not produced in the presence of kifunensine, i.e., have forced high-mannose glycosylation (Doores et al. 2010), and it remains unclear how this barrier can be overcome. A similar approach using synthesized strand C peptides bearing Man$_5$ and Man$_3$ glycans has been proposed for the elicitation of V2-apex directed bnAbs (Alam et al. 2013), but more sophisticated immunization strategies using native trimers will likely be necessary for the induction of bnAbs to this quaternary epitope cluster.

4.3 Challenges an Approaches for the Design of Universal Influenza Vaccines

Heterosubtypic antibodies to influenza virus are less hypermutated than bnAbs to HIV and display a skewed VH region gene usage. The VH1-69 germline gene, found in heterosubtypic antibodies CR9114, CR6261, F10 and 1.12, encodes for hydrophobic residues Ile53 and Phe54, which, in the case of CR6261 and CR9114, interact with a hydrophobic pocket adjacent to helix A. The second most predominant class of heterosubtypic antibodies uses the VH3-30 germline gene for the heavy chain in combination with VK4-1 and Vk-1-12 light chains.

As it has been shown for some HIV bnAbs that fail to bind to recombinant Env, but were able to trigger B cell signaling when expressed as IgM on B cells (Ota et al. 2012), also HA-specific germline-reverted VH1-69 antibodies fail to bind to soluble HA at concentrations as high as 100 µg/ml (Lingwood et al. 2012). For VH1-69-encoded antibodies, a triad of residues in the heavy chain are key for breadth: two residues in HCDR2, i.e., a hydrophobic residue at position 53 and Phe54 interacting with one hydrophobic pocket, in combination with a properly positioned Tyr residue at position 98 $\pm$ 1 being inserted into an adjacent hydrophobic pockets. HnAbs can evolve from the VH1-69 germline with as little as one V-region substitution when combined with a properly positioned HCDR3 Tyrosine: Replacement of Ile 53 with a smaller small amino acid, such as Ser, Ala or Gly, allows the germline-encoded Phe 54 and the HCDR3 Tyrosine to be inserted into adjacent hydrophobic pockets and establish the key contacts (Avnir et al. 2014).

In contrast to HIV, development of a pan flu vaccine mostly involves common denominator immunizations (c.f, Fig. 7). As a matter of fact, the seasonal influenza vaccinations to some extent represent both, a cocktail and a sequential immunization strategy. Current influenza vaccines are tri- or quadrivalent and contain two or three IAVs (one H3N3, one or two H1N1) and one influenza B virus. The temporal modifications of the vaccine formulation due to antigenic shift also represents a sequential immunization strategy. While this strategy does not induce protective heterosubtypic immunity, i.e., people still get infected with shifted or drifted strains, there is evidence that (repetitive) vaccination indeed broadens the antibody response (Kohler et al. 2014; Henry Dunand et al. 2015). Experimentally, this approach has further been refined into an immunization strategy in which the relatively conserved stem has been grafted with globular HA heads derived from different IAV subtypes (Fig. 8, right). Using iterations of these grafted antigens, antibodies were induced that could bind different subtypes of HA, yet failed to

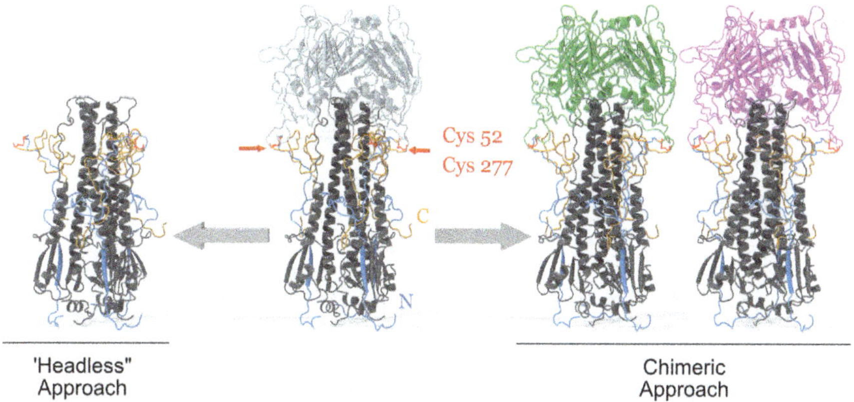

Fig. 8 Immunogen design strategies for eliciting heterosubtypic antibodies. The structure of the H3 HA from A/Aichi/2/1968 H3N2 (PDB 2HMG) is shown with HA2 in gray, and the globular head of HA1 in white. The N-terminal ascending strand of HA1 is depicted in blue and the descending C-terminal end of HA1 in orange, both are linked to each other by the indicated disulfide bridge. The two primary immunogen strategies are (i) 'headless' in which the apical globular portion of HA has been removed and (ii) the chimeric approach in which the globular HA1 head is exchanged with that of different isolates/subtypes (purple and green)

neutralize the corresponding viruses. Despite the lack of heterosubtypic neutralization, some protection could be induced in mice immunized with this strategy (Krammer et al. 2013). However, mice represent a suboptimal animal model for IAV, as protection from lethal infections is relatively easily achieved in mice.

Unlike HIV, where vaccinees are expected to be immunologically naïve, the majority of the recipients of a universal influenza virus vaccine will have preexisting immunity to influenza. This preexisting immunity imposes several challenges: First, there will be plenty of memory B cells specific for the immunodominant epitopes, and it will therefore be more challenging to avoid responses against unwanted immunodominant epitopes. Second, a phenomenon described as Original Antigenic Sin (de St Groth and Webster 1966; Gostic et al. 2016), may contribute another factor that possibly impairs the development of an universal vaccine to influenza virus: It describes the fact that immunization or infection with a new influenza variant rather boosts memory responses to the previous 'original' variants than priming responses to the new antigenic variant. Besides this classical Original Antigenic Sin, it has recently also been described that the first virus encountered will result in a livelong imprinting that favors immune responses to viruses belonging to the same phylogenetic group as the initially encountered virus (Kohler et al. 2014; Gostic et al. 2016). This imprinting, together with cross-reactivity of H3 mAbs to H7 (Henry Dunand et al. 2015), has been postulated to be responsible for the age-related differences in lethality following zoonotic infection with H5N1 and H7N9 HPAI viruses, in that older individuals that have first been exposed to H1N1-expressing viruses cope better with H5N1 infections (both phylogenetic group 1) while younger individuals, more likely to be exposed to H3N2 viruses, cope better with H7N9 zoonotic infections (both phylogenetic group 2) (Gostic et al. 2016).

Although the impact of original antigenic sin on the development of a universal vaccine to influenza is controversial, issues with preexisting immunity can only be avoided if a universal influenza virus vaccine is applied very early in life, and before the virus was encountered for the first time.

4.3.1 hnAb Engineering Approaches to Induce Heterosubtypic Antibodies

There are two major protein-engineering approaches to design immunogens capable of eliciting heterosubtypic antibodies to influenza viruses. These take advantage of a naturally occurring disulfide-bridge formed between Cys 52 and Cys 277, that links the ascending N-terminal with the descending C-terminal end of HA1 (Fig. 8).

The first approach, referred to as 'headless-HA' or 'mini-HA,' replaces the globular head domain with a short linker between these two cysteines. However, to prevent the meta-stable, spring-loaded fusion machinery contained in the stem from folding into the enthalpy-preferred post-fusion six-helix bundle, extensive stabilizing mutations have to be introduced in HA2 to preserve the pre-fusion conformation (Sagawa et al. 1996; Steel et al. 2010; Impagliazzo et al. 2015; van der Lubbe et al. 2018). This approach has the advantage that all of the variable head epitopes are missing but this comes at the expense that epitopes normally buried in the trimer become exposed. In case of HIV BG505 SOSIP.664 trimer immunization, for instance, the neo-epitopes on the bottom of the trimer have proven to be very highly immunogenic (Bianchi et al. 2018), and it will remain to be seen to which extent this will be an issue.

The second approach is a classical 'common denominator' strategy and takes advantage of chimeric proteins in which the stem is kept constant while the HA1 head is exchanged between immunizations (Krammer et al. 2013, 2014; Nachbagauer et al. 2016, 2017; Graham et al. 2019). Such chimeric HA proteins have been shown to display a more open conformation of the head, but this did not affect binding to HA-specific antibodies (Tran et al. 2016).

Both approaches were able to induce heterosubtyic nnAbs that provided protection in mice (Krammer et al. 2013, 2014; Nachbagauer et al. 2016) and ferrets (Nachbagauer et al. 2017) but so far none of these attempts succeeded at inducing heterosubtypic neutralization. The observed protection in mice is most likely mediated by non-neutralizing antibodies via FcγR effector functions. However, it is rather unlikely that similar protection by nnAbs will be achieved in humans, as most humans possess average ELISA titers of 1:1200 to heterologous human and 1:70–100 to nonhuman IAV subtypes (Kohler et al. 2014); yet most people are susceptible to seasonal (or zoonotic) influenza viruses.

4.3.2 Alternative Approaches to Induce Heterosubtypic Antibodies

Besides these two main strategies, several other approaches have been described. These include the removal of glycans (Liu et al. 2016), immunization with mRNA in liposomes (Pardi et al. 2018), antigen dilution (Kanekiyo et al. 2019), and expression of a HA/CD40L fusion protein by recombinant adenoviruses (Fan et al. 2014). The last approach has been the only one to report successful elicitation of heterosubtypic neutralization (Fan et al. 2014). Besides these approaches aiming for the HA surface protein, alternative strategies targeting the relatively conserved M2 protein ectodomain (Jegerlehner et al. 2004; Lo et al. 2008; Rao et al. 2010; Simhadri et al. 2015), and the nucleoprotein have been described (Lo et al. 2008; Rao et al. 2010), albeit with limited success.

Thus, for influenza, a number of strategies have been developed that successfully elicited antibodies that were able to recognize multiple subtypes by ELISA, and that could protect mice from lethal heterologous infections. However, with the exception of a single report, none of the published approaches successfully elicited heterosubtypic neutralization despite detection of stem-specific antibodies. It can be assumed that although these antibodies recognize conserved stem epitopes, their binding does not prevent the structural changes required for neutralizing activity. It is therefore likely that a refinement of the current common denominator strategies for influenza viruses with shepherding approaches, in order to guide the antibody fine-recognition from binding to neutralization, will be successful at eliciting heterosubtypic neutralization. Moreover, passive immunization experiments using stem-specific heterosubtypic antibodies indicate that these antibodies heavily rely on Fc-effector functions for protection, suggesting that the isotype profile of the vaccine response may be of importance for such an IAV vaccine (DiLillo et al. 2014, 2016; Asthagiri Arunkumar et al. 2019; Vanderven and Kent 2020).

In contrast to HIV, where the germline-encoded human antibody repertoire appears to be ill-equipped to recognize many bnAb epitopes, it was found that the presence of the human VH1-69 germline gene in combination with the human D and J elements was sufficient to enable transgenic mice to generate stem-specific antibodies (Sangesland et al. 2019), it therefore might not be necessary to design germline-trargeting immunogens. However, since access to stem-specific epitopes might be restricted on the densely packed surface of influenza virion (Fig. 2), we tested the influence of this possible restraint in an animal experiment (Bianchi, Hangartner, unpublished): in order to expose the conserved stem epitopes more prominently while at the same time occluding access to the variable head, recombinant HA protein was teathered head-down to magnetite beads that were then used for immunization. Two HA proteins were chosen for this approach: first, HA from A/Japan/305/1957 (H2N2), as a representative of phylogenetic group 1, and because of its antigenic properties. Both inverted proteins were very efficient at isolating hnAbs by phage display, including pan-IAV neutralizing mAb 1.12 that was isolated from a normal donor (Wyrzucki et al. 2014, 2015). In contrast to H1N1, where five distinct antigenic sites on the HA globular head have been identified (Gerhard et al. 1981; Caton et al. 1982), and H3N2 IAVs (Wiley et al. 1981; Daniels et al. 1983), H2N2 viruses were shown to have an additional sixth antigenic domain in the stem region (Okuno et al. 1993; Tsuchiya et al. 2001). Therefore, as a second immunogen, HA from A/FPV/Bratislava/1979 (H7N7) was selected, as it belongs to phylogenetic group 2, and also due to its evolutionary distance to the circulating human H3 subtype. Moreover, both H2 and H7 represent a current public health threat. However, although the inverted immunogens readily produced cross-reactive and cross-protective antibodies in mice, they failed to induce heterosubtypic neutralization (Bianchi, Hangartner, unpublished). Hence, although epitope-geography appears to play a role, by itself, it fails to explain immunoquiescence of the hnAb stem epitope.

5 Conclusions

Although immense progress has been made over the past years, universal vaccines to highly variable viruses are still out of reach for now. The primary problems identified are the poor immunogenicity of the shared and conserved epitopes, the immunodominance of undesired variable epitopes, as well as difficulties with constructing stable immunogens that faithfully mimic viral spikes proteins immunologically. Meanwhile, technologies have been developed to not only express stable mimetics of the viral spike proteins, but also to modify them, in order to better engage germline-encoded bnAb/hnAb precursor B cells, or even to generate dedicated antigens designed to best engage germline lineages of broadly neutralizing antibodies. The latter appears to be necessary, as the human germline-encoded B cell repertoire struggles with the recognition of these epitopes. These difficulties reflect the viral evolution to avoid responses to the viral 'Archilles' heel' of the respective surface proteins. The underlying molecular mechanisms for the different immunogenicities of the various epitopes are complex and not yet well understood.

Current research and clinical trials employing both common denominator and sheparding approaches, as well as unmodified spike proteins, however, will provide valuable insight into these issues and will help to inform iterative development of diverse vaccination approaches and immunogens. Almost as a side product, these efforts are also providing profound and fundamental insights into B cell biology and humoral antibody responses more generally along the way.

References

Alam SM, Dennison SM, Aussedat B et al (2013) Recognition of synthetic glycopeptides by HIV-1 broadly neutralizing antibodies and their unmutated ancestors. Proc Natl Acad Sci USA 110:18214–18219. https://doi.org/10.1073/pnas.1317855110

Altman MO, Angel M, Košík I et al (2017) Glycan clock forecasts human influenza A virus Hemagglutinin evolution. bioRxiv 163345. https://doi.org/10.1101/163345

Andrabi R, Voss JE, Liang C-H et al (2015) Identification of common features in prototype broadly neutralizing antibodies to HIV envelope V2 apex to facilitate vaccine design. Immunity 43:959–973. https://doi.org/10.1016/j.immuni.2015.10.014

Angeletti D, Gibbs JS, Angel M et al (2017) Defining B cell immunodominance to viruses. Nat Immunol 18:456–463. https://doi.org/10.1038/ni.3680

Asthagiri Arunkumar G, Ioannou A, Wohlbold TJ et al (2019) Broadly cross-reactive, nonneutralizing antibodies against influenza B virus hemagglutinin demonstrate effector function-dependent protection against lethal viral challenge in mice. J Virol 93:45. https://doi.org/10.1128/JVI.01696-18

Avnir Y, Tallarico AS, Zhu Q et al (2014) Molecular signatures of hemagglutinin stem-directed heterosubtypic human neutralizing antibodies against influenza A viruses. PLoS Pathog 10: e1004103. https://doi.org/10.1371/journal.ppat.1004103

Bachmann MF, Ecabert B, Kopf M (1999) Influenza virus: a novel method to assess viral and neutralizing antibody titers in vitro. J Immunol Methods 225:105–111. https://doi.org/10.1016/S0022-1759(99)00034-4

Barbas CF, Bjorling E, Chiodi F et al (1992) Recombinant human Fab fragments neutralize human type 1 immunodeficiency virus in vitro. Proc Natl Acad Sci USA 89:9339–9343. https://doi.org/10.1073/pnas.89.19.9339

Barbas CF, Hu D, Dunlop N et al (1994) In vitro evolution of a neutralizing human antibody to human immunodeficiency virus type 1 to enhance affinity and broaden strain cross-reactivity. Proc Natl Acad Sci USA 91:3809–3813

Bean WJ, Gorman OT, Chambers TM, Kawaoka Y (1992) Evolution and ecology of influenza A viruses. Microbiol Rev 56:152–179. https://doi.org/10.1007/978-3-642-77011-1_6

Bianchi M, Turner HL, Nogal B et al (2018) Electron-microscopy-based epitope mapping defines specificities of polyclonal antibodies elicited during HIV-1 BG505 envelope trimer immunization. Immunity 49:288–300.e8. https://doi.org/10.1016/j.immuni.2018.07.009

Binley JM, Lybarger EA, Crooks ET et al (2008) Profiling the specificity of neutralizing antibodies in a large panel of plasmas from patients chronically infected with human immunodeficiency virus type 1 subtypes B and C. J Virol 82:11651–11668. https://doi.org/10.1128/JVI.01762-08

Binley JM, Sanders RW, Clas B et al (2000) A recombinant human immunodeficiency virus type 1 envelope glycoprotein complex stabilized by an intermolecular disulfide bond between the gp120 and gp41 subunits is an antigenic mimic of the trimeric virion-associated structure. J Virol 74:627–643. https://doi.org/10.1128/JVI.74.2.627-643.2000

Binley JM, Wrin T, Korber B et al (2004) Comprehensive cross-clade neutralization analysis of a panel of anti-human immunodeficiency virus type 1 monoclonal antibodies. J Virol 78:13232–13252. https://doi.org/10.1128/JVI.78.23.13232-13252.2004

Blattner C, Lee JH, Sliepen K et al (2014) Structural delineation of a quaternary, cleavage-dependent epitope at the gp41-gp120 interface on intact HIV-1 Env trimers. Immunity 40:669–680. https://doi.org/10.1016/j.immuni.2014.04.008

Bonomelli C, Doores KJ, Dunlop DC et al (2011) The glycan shield of HIV is predominantly oligomannose independently of production system or viral clade. PLoS ONE 6:e23521. https://doi.org/10.1371/journal.pone.0023521

Bonsignori M, Kreider EF, Fera D et al (2017a) Staged induction of HIV-1 glycan-dependent broadly neutralizing antibodies. Sci Transl Med 9:eaai7514. https://doi.org/10.1126/scitranslmed.aai7514

Bonsignori M, Liao H-X, Gao F et al (2017b) Antibody-virus co-evolution in HIV infection: paths for HIV vaccine development. Immunol Rev 275:145–160. https://doi.org/10.1111/imr.12509

Bonsignori M, Zhou T, Sheng Z et al (2016) Maturation pathway from germline to broad HIV-1 neutralizer of a CD4-mimic antibody. Cell 165:449–463. https://doi.org/10.1016/j.cell.2016.02.022

Borrow P, Lewicki H, Hahn BH et al (1994) Virus-specific CD8 + cytotoxic T-lymphocyte activity associated with control of viremia in primary human immunodeficiency virus type 1 infection. J Virol 68:6103–6110. https://doi.org/10.5353/th_b4163407

Both GW, Sleigh MJ, Cox NJ, Kendal AP (1983) Antigenic drift in influenza virus H3 hemagglutinin from 1968 to 1980: multiple evolutionary pathways and sequential amino acid changes at key antigenic sites. J Virol 48:52–60

Böttcher E, Matrosovich T, Beyerle M et al (2006) Proteolytic activation of influenza viruses by serine proteases TMPRSS2 and HAT from human airway epithelium. J Virol 80:9896–9898. https://doi.org/10.1128/JVI.01118-06

Brandenberg OF, Magnus C, Rusert P et al (2015) Different infectivity of HIV-1 strains is linked to number of envelope trimers required for entry. PLoS Pathog 11:e1004595. https://doi.org/10.1371/journal.ppat.1004595

Briney B, Sok D, Jardine JG et al (2016) Tailored immunogens direct affinity maturation toward HIV neutralizing antibodies. Cell 166:1459–1470.e11. https://doi.org/10.1016/j.cell.2016.08.005

Buchacher A, Predl R, Strutzenberger K et al (1994) Generation of human monoclonal antibodies against HIV-1 proteins; electrofusion and Epstein-Barr virus transformation for peripheral blood lymphocyte immortalization. AIDS Res Human Retroviruses 10:359–369. https://doi.org/10.1089/aid.1994.10.359

Bullough PA, Hughson FM, Wiley DC (1994) Structure of influenza haemagglutinin at the pH of membrane fusion. Nature 371:37–43. https://doi.org/10.1038/371037a0

Burton DR (2017) What are the most powerful immunogen design vaccine strategies? Reverse vaccinology 2.0 shows great promise. Cold Spring Harb Perspect Biol 9:a030262. https://doi.org/10.1101/cshperspect.a030262

Burton DR (2002) Antibodies, viruses and vaccines. Nat Rev Immunol 2:706–713. https://doi.org/10.1038/nri891

Burton DR, Barbas CF, Persson MA et al (1991) A large array of human monoclonal antibodies to type 1 human immunodeficiency virus from combinatorial libraries of asymptomatic seropositive individuals. Proc Natl Acad Sci USA 88:10134–10137

Burton DR, Hangartner L (2016) Broadly neutralizing antibodies to HIV and their role in vaccine design. Annu Rev Immunol 34:635–659. https://doi.org/10.1146/annurev-immunol-041015-055515

Burton DR, Poignard P, Stanfield RL, Wilson IA (2012) Broadly neutralizing antibodies present new prospects to counter highly antigenically diverse viruses. Science 337:183–186. https://doi.org/10.1126/science.1225416

Burton DR, Pyati J, Koduri R et al (1994) Efficient neutralization of primary isolates of HIV-1 by a recombinant human monoclonal antibody. Science 266:1024–1027. https://doi.org/10.1126/science.7973652

Calarese DA, Scanlan CN, Zwick MB et al (2003) Antibody domain exchange is an immunological solution to carbohydrate cluster recognition. Science 300:2065–2071. https://doi.org/10.1126/science.1083182

Calder LJ, Wasilewski S, Berriman JA, Rosenthal PB (2010) Structural organization of a filamentous influenza A virus. Proc Natl Acad Sci USA 107:10685–10690. https://doi.org/10.1073/pnas.1002123107

Cale EM, Gorman J, Radakovich NA et al (2017) Virus-like particles identify an HIV V1V2 apex-binding neutralizing antibody that lacks a protruding loop. Immunity 46:777–791.e10. https://doi.org/10.1016/j.immuni.2017.04.011

Cao L, Diedrich JK, Kulp DW et al (2017) Global site-specific N-glycosylation analysis of HIV envelope glycoprotein. Nat Commun 8:14954. https://doi.org/10.1038/ncomms14954

Cao L, Diedrich JK, Ma Y et al (2018a) Global site-specific analysis of glycoprotein N-glycan processing. Nat Protoc 13:1196–1212. https://doi.org/10.1038/nprot.2018.024

Cao L, Pauthner M, Andrabi R et al (2018b) Differential processing of HIV envelope glycans on the virus and soluble recombinant trimer. Nat Commun 9:3693–14. https://doi.org/10.1038/s41467-018-06121-4

Caton AJ, Brownlee GG, Yewdell JW, Gerhard W (1982) The antigenic structure of the influenza virus A/PR/8/34 hemagglutinin (H1 subtype). Cell 31:417–427

Cattoli G, Milani A, Temperton N et al (2011) Antigenic drift in H5N1 avian influenza virus in poultry is driven by mutations in major antigenic sites of the hemagglutinin molecule analogous to those for human influenza virus. J Virol 85:8718–8724. https://doi.org/10.1128/JVI.02403-10

Conley AJ, Kessler JA, Boots LJ et al (1994) Neutralization of divergent human immunodeficiency virus type 1 variants and primary isolates by IAM-41-2F5, an anti-gp41 human monoclonal antibody. Proc Natl Acad Sci USA 91:3348–3352. https://doi.org/10.1073/pnas.91.8.3348

Corti D, Voss J, Gamblin SJ et al (2011) A neutralizing antibody selected from plasma cells that binds to group 1 and group 2 influenza A hemagglutinins. Science 333:850–856. https://doi.org/10.1126/science.1205669

Coss KP, Vasiljevic S, Pritchard LK et al (2016) HIV-1 glycan density drives the persistence of the mannose patch within an infected individual. J Virol 90:11132–11144. https://doi.org/10.1128/JVI.01542-16

Coudeville L, Bailleux F, Riche B et al (2010) Relationship between haemagglutination-inhibiting antibody titres and clinical protection against influenza: development and application of a bayesian random-effects model. BMC Med Res Methodol 10:18. https://doi.org/10.1186/1471-2288-10-18

Crooks ET, Osawa K, Tong T et al (2017) Effects of partially dismantling the CD4 binding site glycan fence of HIV-1 envelope glycoprotein trimers on neutralizing antibody induction. Virology 505:193–209. https://doi.org/10.1016/j.virol.2017.02.024

Crooks ET, Tong T, Chakrabarti B et al (2015) Vaccine-elicited tier 2 HIV-1 neutralizing antibodies bind to quaternary epitopes involving glycan-deficient patches proximal to the CD4 binding site. PLoS Pathog 11:e1004932. https://doi.org/10.1371/journal.ppat.1004932

Crooks ET, Tong T, Osawa K, Binley JM (2011) Enzyme digests eliminate nonfunctional Env from HIV-1 particle surfaces, leaving native Env trimers intact and viral infectivity unaffected. J Virol 85:5825–5839. https://doi.org/10.1128/JVI.00154-11

Dalgleish AG, Beverley PC, Clapham PR et al (1984) The CD4 (T4) antigen is an essential component of the receptor for the AIDS retrovirus. Nature 312:763–767. https://doi.org/10.1038/312763a0

Daniels RS, Douglas AR, Skehel JJ, Wiley DC (1983) Analyses of the antigenicity of influenza haemagglutinin at the pH optimum for virus-mediated membrane fusion. J Gen Virol 64:1657–1662. https://doi.org/10.1099/0022-1317-64-8-1657

Das SR, Hensley SE, David A et al (2011) Fitness costs limit influenza A virus hemagglutinin glycosylation as an immune evasion strategy. Proc Natl Acad Sci USA 108:E1417–E1422. https://doi.org/10.1073/pnas.1108754108

De Jong JC, Rimmelzwaan GF, Fouchier RA, Osterhaus AD (2000) Influenza virus: a master of metamorphosis. J Infect 40:218–228. https://doi.org/10.1053/jinf.2000.0652

De Jong JC, Smith DJ, Lapedes AS et al (2007) Antigenic and genetic evolution of swine influenza A (H3N2) viruses in Europe. J Virol 81:4315–4322. https://doi.org/10.1128/JVI.02458-06

De Marco D, Clementi N, Mancini N et al (2012) A non-VH1-69 heterosubtypic neutralizing human monoclonal antibody protects mice against H1N1 and H5N1 viruses. PLoS ONE 7:e34415. https://doi.org/10.1371/journal.pone.0034415

de St Groth F, Webster RG (1966) Disquisitions of original antigenic sin. I. Evidence in man. J Exp Med 124:331–345. https://doi.org/10.1084/jem.124.3.331

de Taeye SW, Ozorowski G, Torrents de la Peña A et al (2015) Immunogenicity of stabilized HIV-1 envelope trimers with reduced exposure of non-neutralizing epitopes. Cell 163:1702–1715. https://doi.org/10.1016/j.cell.2015.11.056

Decroly E, Vandenbranden M, Ruysschaert JM et al (1994) The convertases furin and PC1 can both cleave the human immunodeficiency virus (HIV)-1 envelope glycoprotein gp160 into gp120 (HIV-1 SU) and gp41 (HIV-I TM). J Biol Chem 269:12240–12247

DiLillo DJ, Palese P, Wilson PC, Ravetch JV (2016) Broadly neutralizing anti-influenza antibodies require Fc receptor engagement for in vivo protection. J Clin Invest 126:605–610. https://doi.org/10.1172/JCI84428

DiLillo DJ, Tan GS, Palese P (2014) Broadly neutralizing hemagglutinin stalk-specific antibodies require FcγR interactions for protection against influenza virus in vivo. Nat Med 20:143–151. https://doi.org/10.1038/nm.3443

Dintzis HM, Dintzis RZ, Vogelstein B (1976) Molecular determinants of immunogenicity: the immunon model of immune response. Proc Natl Acad Sci USA 73:3671–3675. https://doi.org/10.1073/pnas.73.10.3671

Dobay MP, Dobay A, Bantang J, Mendoza E (2011) How many trimers? Modeling influenza virus fusion yields a minimum aggregate size of six trimers, three of which are fusogenic. Mol Biosyst 7:2741–2749. https://doi.org/10.1039/c1mb05060e

Doores KJ, Bonomelli C, Harvey DJ et al (2010) Envelope glycans of immunodeficiency virions are almost entirely oligomannose antigens. Proc Natl Acad Sci USA 107:13800–13805. https://doi.org/10.1073/pnas.1006498107

Doria-Rose NA, Bhiman JN, Roark RS et al (2015) New member of the V1V2-directed CAP256-VRC26 lineage that shows increased breadth and exceptional potency. J Virol 90:76–91. https://doi.org/10.1128/JVI.01791-15

Doria-Rose NA, Schramm CA, Gorman J et al (2014) Developmental pathway for potent V1V2-directed HIV-neutralizing antibodies. Nature 509:55–62. https://doi.org/10.1038/nature13036

Doyle-Cooper C, Hudson KE, Cooper AB et al (2013) Immune tolerance negatively regulates B cells in knock-in mice expressing broadly neutralizing HIV antibody 4E10. 191:3186–3191. https://doi.org/10.4049/jimmunol.1301285

Dreyfus C, Laursen NS, Kwaks T et al (2012) Highly conserved protective epitopes on influenza B viruses. Science 337:1343–1348. https://doi.org/10.1126/science.1222908

Dubrovskaya V, Guenaga J, de Val N et al (2017) Targeted N-glycan deletion at the receptor-binding site retains HIV Env NFL trimer integrity and accelerates the elicited antibody response. PLoS Pathog 13:e1006614. https://doi.org/10.1371/journal.ppat.1006614

Ekiert DC, Bhabha G, Elsliger M-A et al (2009) Antibody recognition of a highly conserved influenza virus epitope. Science 324:246–251. https://doi.org/10.1126/science.1171491

Ekiert DC, Kashyap AK, Steel J et al (2012) Cross-neutralization of influenza A viruses mediated by a single antibody loop. Nature 489:526–532. https://doi.org/10.1038/nature11414

Escolano A, Steichen JM, Dosenovic P et al (2016) Sequential immunization elicits broadly neutralizing Anti-HIV-1 antibodies in Ig knockin mice. Cell 166:1445–1458.e12. https://doi.org/10.1016/j.cell.2016.07.030

Falkowska E, Ramos A, Feng Y et al (2012) PGV04, an HIV-1 gp120 CD4 binding site antibody, is broad and potent in neutralization but does not induce conformational changes characteristic of CD4. J Virol 86:4394–4403. https://doi.org/10.1128/JVI.06973-11

Falkowska E, Ramos A, Lee JH et al (2014) Broadly neutralizing HIV antibodies define a glycan-dependent epitope on the prefusion conformation of gp41 on cleaved envelope trimers. Immunity 40:657–668. https://doi.org/10.1016/j.immuni.2014.04.009

Fan X, Hashem AM, Chen Z et al (2014) Targeting the HA2 subunit of influenza A virus hemagglutinin via CD40L provides universal protection against diverse subtypes. Mucosal Immunol 8:211–220. https://doi.org/10.1038/mi.2014.59

Fenner F (1993) Smallpox: emergence, global spread, and eradication. Hist Philos Life Sci 15:397–420

Freund NT, Wang H, Scharf L et al (2017) Coexistence of potent HIV-1 broadly neutralizing antibodies and antibody-sensitive viruses in a viremic controller. Sci Transl Med 9:eaal2144. https://doi.org/10.1126/scitranslmed.aal2144

Friesen RHE, Lee PS, Stoop EJM et al (2014) A common solution to group 2 influenza virus neutralization. Proc Natl Acad Sci USA 111:445–450. https://doi.org/10.1073/pnas.1319058110

Gao F, Liao H-X, Kumar A et al (2014) Cooperation of B cell lineages in induction of HIV-1-broadly neutralizing antibodies. Cell 158:481–491. https://doi.org/10.1016/j.cell.2014.06.022

Garces F, Sok D, Kong L et al (2014) Structural evolution of glycan recognition by a family of potent HIV antibodies. Cell 159:69–79. https://doi.org/10.1016/j.cell.2014.09.009

Georgiev IS, Joyce MG, Yang Y et al (2015) Single-chain soluble BG505.SOSIP gp140 trimers as structural and antigenic mimics of mature closed HIV-1 Env. J Virol 89:5318–5329. https://doi.org/10.1128/JVI.03451-14

Gerhard W, Yewdell J, Frankel ME, Webster R (1981) Antigenic structure of influenza virus haemagglutinin defined by hybridoma antibodies. Nature 290:713–717

Gorman J, Soto C, Yang MM et al (2016) Structures of HIV-1 Env V1V2 with broadly neutralizing antibodies reveal commonalities that enable vaccine design. Nat Struct Mol Biol 23:81–90. https://doi.org/10.1038/nsmb.3144

Gostic KM, Ambrose M, Worobey M, Lloyd-Smith JO (2016) Potent protection against H5N1 and H7N9 influenza via childhood hemagglutinin imprinting. Science 354:722–726. https://doi.org/10.1126/science.aag1322

Graham BS, Gilman MSA, McLellan JS (2019) Structure-based vaccine antigen design. Annu Rev Med 70:91–104. https://doi.org/10.1146/annurev-med-121217-094234

Gray ES, Madiga MC, Hermanus T et al (2011) The neutralization breadth of HIV-1 develops incrementally over four years and is associated with CD4 + T cell decline and high viral load during acute infection. J Virol 85:4828–4840. https://doi.org/10.1128/JVI.00198-11

Greene SA, Ahmed J, Datta SD et al (2019) Progress toward polio eradication—worldwide, January 2017–March 2019. MMWR Morb Mortal Wkly Rep 68:458–462. https://doi.org/10.15585/mmwr.mm6820a3

Gristick HB, von Boehmer L, West AP Jr et al (2016) Natively glycosylated HIV-1 Env structure reveals new mode for antibody recognition of the CD4-binding site. Nat Struct Mol Biol 23:1–12. https://doi.org/10.1038/nsmb.3291

Han Q, Williams WB, Saunders KO et al (2017) HIV DNA-adenovirus multiclade envelope vaccine induces gp41 antibody immunodominance in rhesus macaques. J Virol 91:12449. https://doi.org/10.1128/JVI.00923-17

Hangartner L, Zinkernagel RM, Hengartner H (2006) Antiviral antibody responses: the two extremes of a wide spectrum. Nat Rev Immunol 6:231–243. https://doi.org/10.1038/nri1783

Harris A, Cardone G, Winkler DC et al (2006) Influenza virus pleiomorphy characterized by cryoelectron tomography. Proc Natl Acad Sci USA 103:19123–19127. https://doi.org/10.1073/pnas.0607614103

Havenar-Daughton C, Lee JH, Crotty S (2017) Tfh cells and HIV bnAbs, an immunodominance model of the HIV neutralizing antibody generation problem. Immunol Rev 275:49–61. https://doi.org/10.1111/imr.12512

Haynes BF, Fleming J, St Clair EW et al (2005) Cardiolipin polyspecific autoreactivity in two broadly neutralizing HIV-1 antibodies. Science 308:1906–1908. https://doi.org/10.1126/science.1111781

Henry Dunand CJ, Leon PE, Kaur K et al (2015) Preexisting human antibodies neutralize recently emerged H7N9 influenza strains. J Clin Invest 125:1255–1268. https://doi.org/10.1172/JCI74374

Hessell AJ, Hangartner L, Hunter M et al (2007) Fc receptor but not complement binding is important in antibody protection against HIV. Nature 449:101–104. https://doi.org/10.1038/nature06106

Hessell AJ, Rakasz EG, Poignard P et al (2009) Broadly neutralizing human anti-HIV antibody 2G12 is effective in protection against mucosal SHIV challenge even at low serum neutralizing titers. PLoS Pathog 5:e1000433. https://doi.org/10.1371/journal.ppat.1000433

Hessell AJ, Rakasz EG, Tehrani DM et al (2010) Broadly neutralizing monoclonal antibodies 2F5 and 4E10 directed against the human immunodeficiency virus type 1 gp41 membrane-proximal external region protect against mucosal challenge by simian-human immunodeficiency virus SHIVBa-L. J Virol 84:1302–1313. https://doi.org/10.1128/JVI.01272-09

Hoot S, McGuire AT, Cohen KW et al (2013) Recombinant HIV envelope proteins fail to engage germline versions of anti-CD4bs bNAbs. PLoS Pathog 9:e1003106. https://doi.org/10.1371/journal.ppat.1003106

Huang J, Kang BH, Ishida E et al (2016) Identification of a CD4-binding-site antibody to HIV that evolved near-pan neutralization breadth. Immunity 45:1108–1121. https://doi.org/10.1016/j. immuni.2016.10.027

Huang J, Kang BH, Pancera M et al (2014) Broad and potent HIV-1 neutralization by a human antibody that binds the gp41-gp120 interface. Nature 515:138–142. https://doi.org/10.1038/ nature13601

Huang J, Ofek G, Laub L et al (2012a) Broad and potent neutralization of HIV-1 by a gp41-specific human antibody. Nature 491:406–412. https://doi.org/10.1038/nature11544

Huang J-W, Lin W-F, Yang J-M (2012b) Antigenic sites of H1N1 influenza virus hemagglutinin revealed by natural isolates and inhibition assays. Vaccine 30:6327–6337. https://doi.org/10. 1016/j.vaccine.2012.07.079

Ilyushina N, Rudneva I, Gambaryan A et al (2004) Monoclonal antibodies differentially affect the interaction between the hemagglutinin of H9 influenza virus escape mutants and sialic receptors. Virology 329:33–39. https://doi.org/10.1016/j.virol.2004.08.002

Ilyushina NA, Rudneva IA, Shilov AA et al (2005) Postreassortment changes in a model system: HA-NA adjustment in an H3N2 avian-human reassortant influenza virus. Adv Virol 150:1327–1338. https://doi.org/10.1007/s00705-005-0490-4

Impagliazzo A, Milder F, Kuipers H et al (2015) A stable trimeric influenza hemagglutinin stem as a broadly protective immunogen. Science 349:1301–1306. https://doi.org/10.1126/science. aac7263

Irimia A, Sarkar A, Stanfield RL, Wilson IA (2016) Crystallographic identification of lipid as an integral component of the epitope of HIV broadly neutralizing antibody 4E10. Immunity 44:21–31. https://doi.org/10.1016/j.immuni.2015.12.001

Irimia A, Serra AM, Sarkar A et al (2017) Lipid interactions and angle of approach to the HIV-1 viral membrane of broadly neutralizing antibody 10E8: insights for vaccine and therapeutic design. PLoS Pathog 13:e1006212. https://doi.org/10.1371/journal.ppat.1006212

Jardine J, Julien J-P, Menis S et al (2013) Rational HIV immunogen design to target specific germline B cell receptors. Science 340:711–716. https://doi.org/10.1126/science.1234150

Jardine JG, Kulp DW, Havenar-Daughton C et al (2016a) HIV-1 broadly neutralizing antibody precursor B cells revealed by germline-targeting immunogen. Science 351:1458–1463. https:// doi.org/10.1126/science.aad9195

Jardine JG, Ota T, Sok D et al (2015) HIV-1 VACCINES. Priming a broadly neutralizing antibody response to HIV-1 using a germline-targeting immunogen. Science 349:156–161. https://doi. org/10.1126/science.aac5894

Jardine JG, Sok D, Julien J-P et al (2016b) Minimally mutated HIV-1 broadly neutralizing antibodies to guide reductionist vaccine design. PLoS Pathog 12:e1005815. https://doi.org/10. 1371/journal.ppat.1005815

Jegerlehner A, Schmitz N, Storni T, Bachmann MF (2004) Influenza A vaccine based on the extracellular domain of M2: weak protection mediated via antibody-dependent NK cell activity. J Immunol 172:5598–5605. https://doi.org/10.4049/jimmunol.172.9.5598

Julg B, Sok D, Schmidt SD et al (2017) Protective efficacy of broadly neutralizing antibodies with incomplete neutralization activity against simian-human immunodeficiency virus in rhesus monkeys. J Virol 91:e01187–17. https://doi.org/10.1128/jvi.01187-17

Julien J-P, Cupo A, Sok D et al (2013a) Crystal structure of a soluble cleaved HIV-1 envelope trimer. Science 342:1477–1483. https://doi.org/10.1126/science.1245625

Julien J-P, Khayat R, Walker LM et al (2013b) Broadly neutralizing antibody PGT121 allosterically modulates CD4 binding via recognition of the HIV-1 gp120 V3 base and multiple surrounding glycans. PLoS Pathog 9:e1003342. https://doi.org/10.1371/journal.ppat.1003342

Julien J-P, Lee JH, Cupo A et al (2013c) Asymmetric recognition of the HIV-1 trimer by broadly neutralizing antibody PG9. Proc Natl Acad Sci USA 110:4351–4356. https://doi.org/10.1073/ pnas.1217537110

Kallewaard NL, Corti D, Collins PJ et al (2016) Structure and function analysis of an antibody recognizing all influenza A subtypes. Cell 166:596–608. https://doi.org/10.1016/j.cell.2016.05. 073

Kanekiyo M, Joyce MG, Gillespie RA et al (2019) Mosaic nanoparticle display of diverse influenza virus hemagglutinins elicits broad B cell responses. Nat Immunol 20:362–372. https://doi.org/10.1038/s41590-018-0305-x

Karakus U, Thamamongood T, Ciminski K et al (2019) MHC class II proteins mediate cross-species entry of bat influenza viruses. Nature 567:109–112. https://doi.org/10.1038/s41586-019-0955-3

Kaverin NV, Rudneva IA, Govorkova EA et al (2007) Epitope mapping of the hemagglutinin molecule of a highly pathogenic H5N1 influenza virus by using monoclonal antibodies. J Virol 81:12911–12917. https://doi.org/10.1128/JVI.01522-07

Keele BF, Giorgi EE, Salazar-Gonzalez JF et al (2008) Identification and characterization of transmitted and early founder virus envelopes in primary HIV-1 infection. Proc Natl Acad Sci USA 105:7552–7557. https://doi.org/10.1073/pnas.0802203105

Kepler TB, Liao H-X, Alam SM et al (2014) Immunoglobulin gene insertions and deletions in the affinity maturation of HIV-1 broadly reactive neutralizing antibodies. Cell Host Microbe 16:304–313. https://doi.org/10.1016/j.chom.2014.08.006

Kido H, Murakami M, Oba K et al (1999) Cellular proteinases trigger the infectivity of the influenza A and Sendai viruses. Mol Cells 9:235–244

Kim AS, Leaman DP, Zwick MB (2014) Antibody to gp41 MPER alters functional properties of HIV-1 Env without complete neutralization. PLoS Pathog 10:e1004271. https://doi.org/10.1371/journal.ppat.1004271

Klasse PJ, LaBranche CC, Ketas TJ et al (2016) Sequential and simultaneous immunization of rabbits with HIV-1 envelope glycoprotein SOSIP.664 trimers from clades A, B and C. PLoS Pathog 12:e1005864. https://doi.org/10.1371/journal.ppat.1005864

Klatzmann D, Champagne E, Chamaret S et al (1984) T-lymphocyte T4 molecule behaves as the receptor for human retrovirus LAV. Nature 312:767–768. https://doi.org/10.1038/312767a0

Klein F, Gaebler C, Mouquet H et al (2012) Broad neutralization by a combination of antibodies recognizing the CD4 binding site and a new conformational epitope on the HIV-1 envelope protein. J Exp Med 209:1469–1479. https://doi.org/10.1084/jem.20120423

Kohler I, Scherrer AU, Zagordi O et al (2014) Prevalence and predictors for homo- and heterosubtypic antibodies against influenza a virus. Clin Infect Dis 59:1386–1393. https://doi.org/10.1093/cid/ciu660

Kong L, He L, de Val N et al (2016a) Uncleaved prefusion-optimized gp140 trimers derived from analysis of HIV-1 envelope metastability. Nat Commun 7:12040–15. https://doi.org/10.1038/ncomms12040

Kong L, Lee JH, Doores KJ et al (2013) Supersite of immune vulnerability on the glycosylated face of HIV-1 envelope glycoprotein gp120. 20:796–803. https://doi.org/10.1038/nsmb.2594

Kong R, Xu K, Zhou T et al (2016b) Fusion peptide of HIV-1 as a site of vulnerability to neutralizing antibody. Science 352:828–833. https://doi.org/10.1126/science.aae0474

Korber B, Gaschen B, Yusim K et al (2001) Evolutionary and immunological implications of contemporary HIV-1 variation. Br Med Bull 58:19–42

Krammer F, Margine I, Hai R et al (2014) H3 stalk-based chimeric hemagglutinin influenza virus constructs protect mice from H7N9 challenge. J Virol 88:2340–2343. https://doi.org/10.1128/JVI.03183-13

Krammer F, Pica N, Hai R et al (2013) Chimeric hemagglutinin influenza virus vaccine constructs elicit broadly protective stalk-specific antibodies. J Virol 87:6542–6550. https://doi.org/10.1128/JVI.00641-13

Kulp DW, Steichen JM, Pauthner M et al (2017) Structure-based design of native-like HIV-1 envelope trimers to silence non-neutralizing epitopes and eliminate CD4 binding. Nat Commun 8:1655. https://doi.org/10.1038/s41467-017-01549-6

Landais E, Huang X, Havenar-Daughton C et al (2016) Broadly neutralizing antibody responses in a large longitudinal Sub-Saharan hiv primary infection cohort. PLoS Pathog 12:e1005369–1641. https://doi.org/10.1371/journal.ppat.1005369

Lauring AS, Frydman J, Andino R (2013) The role of mutational robustness in RNA virus evolution. Nat Rev Micro 11:327–336. https://doi.org/10.1038/nrmicro3003

Laursen NS, Wilson IA (2013) Broadly neutralizing antibodies against influenza viruses. Antiviral Res 98:476–483. https://doi.org/10.1016/j.antiviral.2013.03.021

Laver WG, Air GM, Ward CW, Dopheide TAA (1979a) Antigenic drift in type A influenza virus: Sequence differences in the hemagglutinin of Hong Kong (H3N2) variants selected with monoclonal hybridoma antibodies. Virology 98:226–237. https://doi.org/10.1016/0042-6822(79)90540-3

Laver WG, Gerhard W, Webster RG et al (1979b) Antigenic drift in type A influenza virus: peptide mapping and antigenic analysis of A/PR/8/34 (HON1) variants selected with monoclonal antibodies. Proc Nat Acad Sci 76:1425–1429. https://doi.org/10.1073/pnas.76.3.1425

Lee JH, Leaman DP, Kim AS et al (2015) Antibodies to a conformational epitope on gp41 neutralize HIV-1 by destabilizing the Env spike. Nat Commun 6:8167. https://doi.org/10.1038/ncomms9167

Lee JH, Ozorowski G, Ward AB (2016) Cryo-EM structure of a native, fully glycosylated, cleaved HIV-1 envelope trimer. Science 351:1043–1048. https://doi.org/10.1126/science.aad2450

Li F, Li C, Revote J et al (2016) GlycoMinestruct: a new bioinformatics tool for highly accurate mapping of the human N-linked and O-linked glycoproteomes by incorporating structural features. Sci Rep 6:34595–16. https://doi.org/10.1038/srep34595

Li M, Gao F, Mascola JR et al (2005) Human immunodeficiency virus type 1 env clones from acute and early subtype B infections for standardized assessments of vaccine-elicited neutralizing antibodies. J Virol 79:10108–10125. https://doi.org/10.1128/JVI.79.16.10108-10125.2005

Li G-M, Chiu C, Wrammert J et al (2012) Pandemic H1N1 influenza vaccine induces a recall response in humans that favors broadly cross-reactive memory B cells. Proc Natl Acad Sci USA 109:9047–9052. https://doi.org/10.1073/pnas.1118979109

Liang P-H, Wang S-K, Wong C-H (2007) Quantitative analysis of carbohydrate-protein interactions using glycan microarrays: determination of surface and solution dissociation constants. J Am Chem Soc 129:11177–11184. https://doi.org/10.1021/ja072931h

Liao H-X, Lynch R, Alam SM et al (2013) Co-evolution of a broadly neutralizing HIV-1 antibody and founder virus. Nature 496:469–476. https://doi.org/10.1038/nature12053

Lingwood D, McTamney PM, Yassine HM et al (2012) Structural and genetic basis for development of broadly neutralizing influenza antibodies. Nature 489:566–570. https://doi.org/110.1038/nature11371

Liu M, Yang G, Wiehe K et al (2014) Polyreactivity and autoreactivity among HIV-1 antibodies. J Virol 89:784–798. https://doi.org/10.1128/JVI.02378-14

Liu W-C, Jan J-T, Huang Y-J et al (2016) Unmasking stem-specific neutralizing epitopes by abolishing N-linked glycosylation sites of influenza virus hemagglutinin proteins for vaccine design. J Virol 90:8496–8508. https://doi.org/10.1128/JVI.00880-16

Lo C-Y, Wu Z, Misplon JA et al (2008) Comparison of vaccines for induction of heterosubtypic immunity to influenza A virus: cold-adapted vaccine versus DNA prime-adenovirus boost strategies. Vaccine 26:2062–2072. https://doi.org/10.1016/j.vaccine.2008.02.047

Lynch RM, Tran L, Louder MK et al (2012) The development of CD4 binding site antibodies during HIV-1 infection. J Virol 86:7588–7595. https://doi.org/10.1128/JVI.00734-12

Lyumkis D, Julien J-P, de Val N et al (2013) Cryo-EM structure of a fully glycosylated soluble cleaved HIV-1 envelope trimer. Science 342:1484–1490. https://doi.org/10.1126/science.1245627

MacLeod DT, Choi NM, Briney B et al (2016) Early antibody lineage diversification and independent limb maturation lead to broad HIV-1 neutralization targeting the Env high-mannose patch. Immunity 44:1215–1226. https://doi.org/10.1016/j.immuni.2016.04.016

Malherbe DC, Doria-Rose NA, Misher L et al (2011) Sequential immunization with a subtype B HIV-1 envelope quasispecies partially mimics the in vivo development of neutralizing antibodies. J Virol 85:5262–5274. https://doi.org/10.1128/JVI.02419-10

Malherbe DC, Pissani F, Sather DN et al (2014) Envelope variants circulating as initial neutralization breadth developed in two HIV-infected subjects stimulate multiclade neutralizing antibodies in rabbits. J Virol 88:12949–12967. https://doi.org/10.1128/JVI.01812-14

Martinez-Murillo P, Tran K, Guenaga J et al (2017) Particulate array of well-ordered HIV clade C Env trimers elicits neutralizing antibodies that display a unique V2 cap approach. Immunity 46:804–817.e7. https://doi.org/10.1016/j.immuni.2017.04.021

Maurer MA, Meyer L, Bianchi M et al (2018) Glycosylation of human IgA directly inhibits influenza A and other sialic-acid-binding viruses. Cell Rep 23:90–99. https://doi.org/10.1016/j.celrep.2018.03.027

McCoy LE, Burton DR (2017) Identification and specificity of broadly neutralizing antibodies against HIV. Immunol Rev 275:11–20. https://doi.org/10.1111/imr.12484

McCoy LE, Falkowska E, Doores KJ et al (2015) Incomplete neutralization and deviation from sigmoidal neutralization curves for HIV broadly neutralizing monoclonal antibodies. PLoS Pathog 11:e1005110. https://doi.org/10.1371/journal.ppat.1005110

McCoy LE, McKnight A (2017) Lessons learned from humoral responses of HIV patients. Curr Opin HIV AIDS 12:195–202. https://doi.org/10.1097/COH.0000000000000361

McCoy LE, van Gils MJ, Ozorowski G et al (2016) Holes in the glycan shield of the native HIV envelope are a target of trimer-elicited neutralizing antibodies. Cell Rep 16:2327–2338. https://doi.org/10.1016/j.celrep.2016.07.074

McGuire AT, Hoot S, Dreyer AM et al (2013) Engineering HIV envelope protein to activate germline B cell receptors of broadly neutralizing anti-CD4 binding site antibodies. J Exp Med 210:655–663. https://doi.org/10.1084/jem.20122824

McLellan JS, Pancera M, Carrico C et al (2011) Structure of HIV-1 gp120 V1/V2 domain with broadly neutralizing antibody PG9. Nature 480:336–343. https://doi.org/10.1038/nature10696

Medina-Ramírez M, Garces F, Escolano A et al (2017) Design and crystal structure of a native-like HIV-1 envelope trimer that engages multiple broadly neutralizing antibody precursors in vivo. J Exp Med 214:2573–2590. https://doi.org/10.1084/jem.20161160

Moldt B, Rakasz EG, Schultz N et al (2012) Highly potent HIV-specific antibody neutralization in vitro translates into effective protection against mucosal SHIV challenge in vivo. Proc Natl Acad Sci USA 109:18921–18925. https://doi.org/10.1073/pnas.1214785109

Moore PL, Crooks ET, Porter L et al (2006) Nature of nonfunctional envelope proteins on the surface of human immunodeficiency virus type 1. J Virol 80:2515–2528. https://doi.org/10.1128/JVI.80.5.2515-2528.2006

Moore PL, Gray ES, Wibmer CK et al (2012) Evolution of an HIV glycan-dependent broadly neutralizing antibody epitope through immune escape. Nat Med 18:1688–1692. https://doi.org/10.1038/nm.2985

Mouquet H, Scharf L, Euler Z et al (2012) Complex-type N-glycan recognition by potent broadly neutralizing HIV antibodies. Proc Natl Acad Sci USA 109:E3268–E3277. https://doi.org/10.1073/pnas.1217207109

Mouquet H, Scheid JF, Zoller MJ et al (2010) Polyreactivity increases the apparent affinity of anti-HIV antibodies by heteroligation. Nature 467:591–595. https://doi.org/10.1038/nature09385

Muñoz-Barroso I, Salzwedel K, Hunter E, Blumenthal R (1999) Role of the membrane-proximal domain in the initial stages of human immunodeficiency virus type 1 envelope glycoprotein-mediated membrane fusion. J Virol 73:6089–6092

Murin CD, Julien J-P, Sok D et al (2014) Structure of 2G12 Fab2 in complex with soluble and fully glycosylated HIV-1 Env by negative-stain single-particle electron microscopy. J Virol 88:10177–10188. https://doi.org/10.1128/JVI.01229-14

Nachbagauer R, Kinzler D, Choi A et al (2016) A chimeric haemagglutinin-based influenza split virion vaccine adjuvanted with AS03 induces protective stalk-reactive antibodies in mice. NPJ Vaccines 1:153. https://doi.org/10.1038/npjvaccines.2016.15

Nachbagauer R, Liu W-C, Choi A et al (2017) A universal influenza virus vaccine candidate confers protection against pandemic H1N1 infection in preclinical ferret studies. NPJ Vaccines 2:26. https://doi.org/10.1038/s41541-017-0026-4

Nakajima K, Desselberger U, Palese P (1978) Recent human influenza A (H1N1) viruses are closely related genetically to strains isolated in 1950. Nature 274:334–339

Nogal B, Bianchi M, Cottrell CA et al (2020) Mapping polyclonal antibody responses in non-human primates vaccinated with HIV Env trimer subunit vaccines. Cell Rep 30:3755–3765.e7. https://doi.org/10.1016/j.celrep.2020.02.061

Okuno Y, Isegawa Y, Sasao F, Ueda S (1993) A common neutralizing epitope conserved between the hemagglutinins of influenza A virus H1 and H2 strains. J Virol 67:2552–2558

Ota T, Doyle-Cooper C, Cooper AB et al (2012) Anti-HIV B cell lines as candidate vaccine biosensors. J Immunol 189:4816–4824. https://doi.org/10.4049/jimmunol.1202165

Pancera M, Shahzad-Ul-Hussan S, Doria-Rose NA et al (2013) Structural basis for diverse N-glycan recognition by HIV-1-neutralizing V1-V2-directed antibody PG16. 20:804–813. https://doi.org/10.1038/nsmb.2600

Pancera M, Zhou T, Druz A et al (2014) Structure and immune recognition of trimeric pre-fusion HIV-1 Env. Nature 514:455–461. https://doi.org/10.1038/nature13808

Panico M, Bouché L, Binet D et al (2016) Mapping the complete glycoproteome of virion-derived HIV-1 gp120 provides insights into broadly neutralizing antibody binding. Sci Rep 6:32956–32917. https://doi.org/10.1038/srep32956

Pardi N, Parkhouse K, Kirkpatrick E et al (2018) Nucleoside-modified mRNA immunization elicits influenza virus hemagglutinin stalk-specific antibodies. Nat Commun 9:1–12. https://doi.org/10.1038/s41467-018-05482-0

Parren PW, Marx PA, Hessell AJ et al (2001) Antibody protects macaques against vaginal challenge with a pathogenic R5 simian/human immunodeficiency virus at serum levels giving complete neutralization in vitro. J Virol 75:8340–8347. https://doi.org/10.1128/JVI.75.17.8340-8347.2001

Parrish CR, Murcia PR, Holmes EC (2015) Influenza virus reservoirs and intermediate hosts: dogs, horses, and new possibilities for influenza virus exposure of humans. J Virol 89:2990–2994. https://doi.org/10.1128/JVI.03146-14

Pauthner M, Havenar-Daughton C, Sok D et al (2017) Elicitation of robust tier 2 neutralizing antibody responses in nonhuman primates by HIV envelope trimer immunization using optimized approaches. Immunity 46:1073–1088.e6. https://doi.org/10.1016/j.immuni.2017.05.007

Pauthner MG, Nkolola JP, Havenar-Daughton C et al (2019) Vaccine-induced protection from homologous tier 2 SHIV challenge in nonhuman primates depends on serum-neutralizing antibody titers. Immunity 50:241–252.e6. https://doi.org/10.1016/j.immuni.2018.11.011

Peacock T, Reddy K, James J et al (2016) Antigenic mapping of an H9N2 avian influenza virus reveals two discrete antigenic sites and a novel mechanism of immune escape. Sci Rep 6:18745. https://doi.org/10.1038/srep18745

Pejchal R, Doores KJ, Walker LM et al (2011) A potent and broad neutralizing antibody recognizes and penetrates the HIV glycan shield. Science 334:1097–1103. https://doi.org/10.1126/science.1213256

Plotkin SA (2010) Correlates of protection induced by vaccination. Clin Vaccine Immunol 17:1055–1065. https://doi.org/10.1128/CVI.00131-10

Plotkin SA (2013) Commentary: Mumps vaccines: do we need a new one? Pediatr Infect Dis J 32:381–382. https://doi.org/10.1097/INF.0b013e3182809dda

Poignard P, Moulard M, Golez E et al (2003) Heterogeneity of envelope molecules expressed on primary human immunodeficiency virus type 1 particles as probed by the binding of neutralizing and nonneutralizing antibodies. J Virol 77:353–365. https://doi.org/10.1128/JVI.77.1.353-365.2003

Pritchard LK, Spencer DIR, Royle L et al (2015a) Glycan clustering stabilizes the mannose patch of HIV-1 and preserves vulnerability to broadly neutralizing antibodies. Nat Commun 6:7479. https://doi.org/10.1038/ncomms8479

Pritchard LK, Vasiljevic S, Ozorowski G et al (2015b) Structural constraints determine the glycosylation of HIV-1 envelope trimers. Cell Rep 11:1604–1613. https://doi.org/10.1016/j.celrep.2015.05.017

Rao SS, Kong W-P, Wei C-J et al (2010) Comparative efficacy of hemagglutinin, nucleoprotein, and matrix 2 protein gene-based vaccination against H5N1 influenza in mouse and ferret. PLoS ONE 5:e9812. https://doi.org/10.1371/journal.pone.0009812

Rawson JMO, Clouser CL, Mansky LM (2016) Rapid Determination of HIV-1 Mutant Frequencies and Mutation Spectra Using an mCherry/EGFP Dual-Reporter Viral Vector. Methods Mol Biol 1354:71–88. https://doi.org/10.1007/978-1-4939-3046-3_6

Ringe RP, Ozorowski G, Rantalainen K et al (2017) Reducing V3 antigenicity and immunogenicity on soluble, native-like HIV-1 Env SOSIP trimers. J Virol 91:e00677–17. https://doi.org/10.1128/jvi.00677-17

Roach JC, Glusman G, Smit AFA et al (2010) Analysis of genetic inheritance in a family quartet by whole-genome sequencing. Science 328:636–639. https://doi.org/10.1126/science.1186802

Roberts PC, Garten W, Klenk HD (1993) Role of conserved glycosylation sites in maturation and transport of influenza A virus hemagglutinin. J Virol 67:3048–3060

Sagawa H, Ohshima A, Kato I et al (1996) The immunological activity of a deletion mutant of influenza virus haemagglutinin lacking the globular region. J Gen Virol 77(Pt 7):1483–1487. https://doi.org/10.1099/0022-1317-77-7-1483

Salzwedel K, West JT, Hunter E (1999) A conserved tryptophan-rich motif in the membrane-proximal region of the human immunodeficiency virus type 1 gp41 ectodomain is important for Env-mediated fusion and virus infectivity. J Virol 73:2469–2480

Sanders RW, Derking R, Cupo A et al (2013) A next-generation cleaved, soluble HIV-1 Env trimer, BG505 SOSIP.664 gp140, expresses multiple epitopes for broadly neutralizing but not non-neutralizing antibodies. PLoS Pathog 9:e1003618. https://doi.org/10.1371/journal.ppat.1003618

Sanders RW, Moore JP (2017) Native-like Env trimers as a platform for HIV-1 vaccine design. Immunol Rev 275:161–182. https://doi.org/10.1111/imr.12481

Sanders RW, van Gils MJ, Derking R et al (2015) HIV-1 VACCINES. HIV-1 neutralizing antibodies induced by native-like envelope trimers. Science 349:aac4223–aac4223. https://doi.org/10.1126/science.aac4223

Sanders RW, Vesanen M, Schuelke N et al (2002) Stabilization of the soluble, cleaved, trimeric form of the envelope glycoprotein complex of human immunodeficiency virus type 1. J Virol 76:8875–8889

Sangesland M, Ronsard L, Kazer SW et al (2019) Germline-encoded affinity for cognate antigen enables vaccine amplification of a human broadly neutralizing response against influenza virus. Immunity 51:735–749.e8. https://doi.org/10.1016/j.immuni.2019.09.001

Saunders KO, Nicely NI, Wiehe K et al (2017) Vaccine elicitation of high mannose-dependent neutralizing antibodies against the V3-glycan broadly neutralizing epitope in nonhuman primates. Cell Reports 18:2175–2188. https://doi.org/10.1016/j.celrep.2017.02.003

Sanjuán R, Moya A, Elena SF (2004) The distribution of fitness effects caused by single-nucleotide substitutions in an RNA virus. Proc Nat Acad Sci 101:8396–8401. https://doi.org/10.1073/pnas.0400146101

Sanjuán R, Nebot MR, Chirico N et al (2010) Viral mutation rates. J Virol 84:9733–9748. https://doi.org/10.1128/JVI.00694-10

Sauter NK, Bednarski MD, Wurzburg BA et al (1989) Hemagglutinins from two influenza virus variants bind to sialic acid derivatives with millimolar dissociation constants: a 500-MHz proton nuclear magnetic resonance study. Biochemistry 28:8388–8396. https://doi.org/10.1021/bi00447a018

Sautto GA, Kirchenbaum GA, Ross TM (2018) Towards a universal influenza vaccine: different approaches for one goal. Virol J 15:17. https://doi.org/10.1186/s12985-017-0918-y

Scanlan CN, Pantophlet R, Wormald MR et al (2002) The broadly neutralizing anti-human immunodeficiency virus type 1 antibody 2G12 recognizes a cluster of alpha1– > 2 mannose residues on the outer face of gp120. J Virol 76:7306–7321. https://doi.org/10.1128/JVI.76.14.7306-7321.2002

Scanlan CN, Ritchie GE, Baruah K et al (2007) Inhibition of mammalian glycan biosynthesis produces non-self antigens for a broadly neutralising, HIV-1 specific antibody. J Mol Biology 372:16–22. https://doi.org/10.1016/j.jmb.2007.06.027

Scharf L, Scheid JF, Lee JH et al (2014) Antibody 8ANC195 reveals a site of broad vulnerability on the HIV-1 envelope spike. Cell Rep 7:785–795. https://doi.org/10.1016/j.celrep.2014.04.001

Scharf L, Wang H, Gao H et al (2015) Broadly neutralizing antibody 8ANC195 recognizes closed and open states of HIV-1 Env. Cell 162:1379–1390. https://doi.org/10.1016/j.cell.2015.08.035

Scheid JF, Mouquet H, Feldhahn N et al (2009) Broad diversity of neutralizing antibodies isolated from memory B cells in HIV-infected individuals. Nature 458:636–640. https://doi.org/10.1038/nature07930

Scheid JF, Mouquet H, Ueberheide B et al (2011) Sequence and structural convergence of broad and potent HIV antibodies that mimic CD4 binding. Science 333:1633–1637. https://doi.org/10.1126/science.1207227

Scholtissek C, Stech J, Krauss S, Webster RG (2002) Cooperation between the hemagglutinin of avian viruses and the matrix protein of human influenza A viruses. J Virol 76:1781–1786

Schoofs T, Barnes CO, Suh-Toma N et al (2019) Broad and potent neutralizing antibodies recognize the silent face of the HIV envelope. Immunity 50:1513–1529.e9. https://doi.org/10.1016/j.immuni.2019.04.014

Sharma SK, de Val N, Bale S et al (2015) Cleavage-independent HIV-1 Env trimers engineered as soluble native spike mimetics for vaccine design. Cell Rep 11:539–550. https://doi.org/10.1016/j.celrep.2015.03.047

Shingai M, Nishimura Y, Klein F et al (2013) Antibody-mediated immunotherapy of macaques chronically infected with SHIV suppresses viraemia. Nature 503:277–280. https://doi.org/10.1038/nature12746

Simek MD, Rida W, Priddy FH et al (2009) Human immunodeficiency virus type 1 elite neutralizers: individuals with broad and potent neutralizing activity identified by using a high-throughput neutralization assay together with an analytical selection algorithm. J Virol 83:7337–7348. https://doi.org/10.1128/JVI.00110-09

Simhadri VR, Dimitrova M, Mariano JL et al (2015) A human Anti-M2 antibody mediates antibody-dependent cell-mediated cytotoxicity (ADCC) and cytokine secretion by resting and cytokine-preactivated natural killer (NK) cells. PLoS ONE 10:e0124677. https://doi.org/10.1371/journal.pone.0124677

Skehel JJ, Wiley DC (2000) Receptor Binding and Membrane Fusion in Virus Entry: The Influenza Hemagglutinin. Annu Rev Biochem 69:531–569. https://doi.org/10.1146/annurev.biochem.69.1.531

Skehel JJ, Bayley PM, Brown EB et al (1982) Changes in the conformation of influenza virus hemagglutinin at the pH optimum of virus-mediated membrane fusion. Proc Natl Acad Sci USA 79:968–972

Sok D, Laserson U, Laserson J et al (2013) The effects of somatic hypermutation on neutralization and binding in the PGT121 family of broadly neutralizing HIV antibodies. PLoS Pathog 9:e1003754. https://doi.org/10.1371/journal.ppat.1003754

Sok D, Doores KJ, Briney B et al (2014a) Promiscuous glycan site recognition by antibodies to the high-mannose patch of gp120 broadens neutralization of HIV. Sci Transl Med 6:236ra63–236ra63. https://doi.org/10.1126/scitranslmed.3008104

Sok D, van Gils MJ, Pauthner M et al (2014b) Recombinant HIV envelope trimer selects for quaternary-dependent antibodies targeting the trimer apex. Proc Natl Acad Sci USA 111:17624–17629. https://doi.org/10.1073/pnas.1415789111

Sok D, Briney B, Jardine JG et al (2016a) Priming HIV-1 broadly neutralizing antibody precursors in human Ig loci transgenic mice. Science 353:1557–1560. https://doi.org/10.1126/science.aah3945

Sok D, Pauthner M, Briney B et al (2016b) A prominent site of antibody vulnerability on HIV envelope incorporates a motif associated with CCR5 binding and its camouflaging glycans. Immunity 45:31–45. https://doi.org/10.1016/j.immuni.2016.06.026

Stamatatos L, Pancera M, McGuire AT (2017) Germline-targeting immunogens. Immunol Rev 275:203–216. https://doi.org/10.1111/imr.12483

Steel J, Lowen AC, Yondola M et al (2010) Influenza virus vaccine based on the conserved hemagglutinin stalk domain. mBio 1:e00018–10–e00018–10. https://doi.org/10.1128/mbio.00018-10

Steichen JM, Kulp DW, Tokatlian T et al (2016) HIV vaccine design to target germline precursors of glycan-dependent broadly neutralizing antibodies. Immunity 45:483–496. https://doi.org/10.1016/j.immuni.2016.08.016

Stiegler G, Kunert R, PURTSCHER M et al (2001) A potent cross-clade neutralizing human monoclonal antibody against a novel epitope on gp41 of human immunodeficiency virus type 1. AIDS Res Hum Retroviruses 17:1757–1765. https://doi.org/10.1089/08892220152741450

Struwe WB, Stuckmann A, Behrens A-J et al (2017) Global N-Glycan Site Occupancy of HIV-1 gp120 by Metabolic Engineering and High-Resolution Intact Mass Spectrometry. ACS Chem Biol 12:357–361. https://doi.org/10.1021/acschembio.6b00854

Sui J, Hwang WC, Perez S et al (2009) Structural and functional bases for broad-spectrum neutralization of avian and human influenza A viruses. Nat Struct Mol Biol 16:265–273. https://doi.org/10.1038/nsmb.1566

Tahara M, Ito Y, Brindley MA et al (2012) Functional and Structural characterization of neutralizing epitopes of measles virus hemagglutinin protein. J Virol 87:666–675. https://doi.org/10.1128/JVI.02033-12

Tomaras GD, Yates NL, Liu P et al (2008) Initial B-cell responses to transmitted human immunodeficiency virus type 1: virion-binding immunoglobulin M (IgM) and IgG antibodies followed by plasma anti-gp41 antibodies with ineffective control of initial viremia. J Virol 82:12449–12463. https://doi.org/10.1128/JVI.01708-08

Tong S, Li Y, Rivailler P et al (2012) A distinct lineage of influenza A virus from bats. Proc Natl Acad Sci USA 109:4269–4274. https://doi.org/10.1073/pnas.1116200109

Tong S, Zhu X, Li Y et al (2013) New World Bats Harbor Diverse Influenza A Viruses. PLoS Pathog 9:e1003657. https://doi.org/10.1371/journal.ppat.1003657

Torrents de la Peña A, Julien J-P, de Taeye SW et al (2017) Improving the immunogenicity of native-like HIV-1 envelope trimers by hyperstabilization. Cell Rep 20:1805–1817. https://doi.org/10.1016/j.celrep.2017.07.077

Tran EEH, Podolsky KA, Bartesaghi A et al (2016) Cryo-electron microscopy structures of chimeric hemagglutinin displayed on a universal influenza vaccine candidate. mBio 7:e00257. https://doi.org/10.1128/mbio.00257-16

Trkola A, Purtscher M, Muster T et al (1996) Human monoclonal antibody 2G12 defines a distinctive neutralization epitope on the gp120 glycoprotein of human immunodeficiency virus type 1. J Virol 70:1100–1108

Tsuchiya E, Sugawara K, Hongo S et al (2001) Antigenic structure of the haemagglutinin of human influenza A/H2N2 virus. 82:2475–2484. https://doi.org/10.1099/0022-1317-82-10-2475

Ugolini S, Mondor I, Sattentau QJ (1999) HIV-1 attachment: another look. Trends in Microbiology 7:144–149. https://doi.org/10.1016/S0966-842X(99)01474-2

van der Lubbe JEM, Verspuij JWA, Huizingh J et al (2018) Mini-HA is superior to full length hemagglutinin immunization in inducing stem-specific antibodies and protection against group 1 influenza virus challenges in mice. Front Immunol 9:2350. https://doi.org/10.3389/fimmu.2018.02350

Vanderven HA, Kent SJ (2020) The protective potential of Fc-mediated antibody functions against influenza virus and other viral pathogens. Immunol Cell Biol imcb.12312–25. https://doi.org/10.1111/imcb.12312

Verkoczy L, Chen Y, Bouton-Verville H et al (2011) Rescue of HIV-1 broad neutralizing antibody-expressing B cells in 2F5 VH x VL knockin mice reveals multiple tolerance controls. 187:3785–3797. https://doi.org/10.4049/jimmunol.1101633

Verkoczy L, Diaz M, Holl TM et al (2010) Autoreactivity in an HIV-1 broadly reactive neutralizing antibody variable region heavy chain induces immunologic tolerance. Proc Natl Acad Sci USA 107:181–186. https://doi.org/10.1073/pnas.0912914107

Voss JE, Andrabi R, McCoy LE et al (2017) Elicitation of neutralizing antibodies targeting the V2 Apex of the HIV envelope trimer in a wild-type animal model. Cell Rep 21:222–235. https://doi.org/10.1016/j.celrep.2017.09.024

Wagner R, Matrosovich M, Klenk H-D (2002) Functional balance between haemagglutinin and neuraminidase in influenza virus infections. Rev Med Virol 12:159–166. https://doi.org/10.1002/rmv.352

Walker LM, Chan-Hui PY, Wagner D et al (2009) Broad and potent neutralizing antibodies from an African donor reveal a new HIV-1 vaccine target. Science 326:285–289. https://doi.org/10.1126/science.1178746

Walker LM, Huber M, Doores KJ et al (2011) Broad neutralization coverage of HIV by multiple highly potent antibodies. Nature 477:466–470. https://doi.org/10.1038/nature10373

Walker LM, Simek MD, Priddy F et al (2010) A limited number of antibody specificities mediate broad and potent serum neutralization in selected HIV-1 infected individuals. PLoS Pathog 6:e1001028. https://doi.org/10.1371/journal.ppat.1001028

Wang G, Wu Y, Zhou T et al (2013) Mapping of the N-linked glycoproteome of human spermatozoa. J Proteome Res 12:5750–5759. https://doi.org/10.1021/pr400753f

Wang S, Mata-Fink J, Kriegsman B et al (2015) Manipulating the selection forces during affinity maturation to generate cross-reactive HIV antibodies. Cell 160:785–797. https://doi.org/10.1016/j.cell.2015.01.027

Wang S-K, Liang P-H, Astronomo RD et al (2008) Targeting the carbohydrates on HIV-1: interaction of oligomannose dendrons with human monoclonal antibody 2G12 and DC-SIGN. Proc Natl Acad Sci USA 105:3690–3695. https://doi.org/10.1073/pnas.0712326105

Wei X, Decker JM, Wang S et al (2003) Antibody neutralization and escape by HIV-1. Nature 422:307–312. https://doi.org/10.1038/nature01470

Wibmer CK, Gorman J, Ozorowski G et al (2017) Structure and recognition of a novel HIV-1 gp120-gp41 interface antibody that caused MPER exposure through viral escape. PLoS Pathog 13:e1006074–1641. https://doi.org/10.1371/journal.ppat.1006074

Wiley DC, Wilson IA, Skehel JJ (1981) Structural identification of the antibody-binding sites of Hong Kong influenza haemagglutinin and their involvement in antigenic variation. Nature 289:373–378. https://doi.org/10.1038/289373a0

Williams LD, Ofek G, Schätzle S et al (2017) Potent and broad HIV-neutralizing antibodies in memory B cells and plasma. Sci Immunol 2:eaal2200. https://doi.org/10.1126/sciimmunol.aal2200

Williams WB, Liao H-X, Moody MA et al (2015) HIV-1 VACCINES. Diversion of HIV-1 vaccine-induced immunity by gp41-microbiota cross-reactive antibodies. Science 349:aab1253–aab1253. https://doi.org/10.1126/science.aab1253

Wood N, Bhattacharya T, Keele BF et al (2009) HIV evolution in early infection: selection pressures, patterns of insertion and deletion, and the impact of APOBEC. PLoS Pathog 5:e1000414. https://doi.org/10.1371/journal.ppat.1000414

Wu NC, Wilson IA (2017) A Perspective on the structural and functional constraints for immune evasion: insights from influenza virus. J Mol Biol 429:2694–2709. https://doi.org/10.1016/j.jmb.2017.06.015

Wu X, Yang Z-Y, Li Y et al (2010) Rational design of envelope identifies broadly neutralizing human monoclonal antibodies to HIV-1. Science 329:856–861. https://doi.org/10.1126/science.1187659

Wyatt R, Kwong PD, Desjardins E et al (1998) The antigenic structure of the HIV gp120 envelope glycoprotein. Nature 393:705–711. https://doi.org/10.1038/31514

Wyrzucki A, Bianchi M, Kohler I et al (2015) Heterosubtypic antibodies to influenza A virus have limited activity against cell-bound virus but are not impaired by strain-specific serum antibodies. J Virol 89:3136–3144. https://doi.org/10.1128/JVI.03069-14

Wyrzucki A, Dreyfus C, Kohler I et al (2014) Alternative recognition of the conserved stem epitope in influenza A virus hemagglutinin by a VH3-30-encoded heterosubtypic antibody. J Virol 88:7083–7092. https://doi.org/10.1128/JVI.00178-14

Xiao X, Chen W, Feng Y et al (2009) Germline-like predecessors of broadly neutralizing antibodies lack measurable binding to HIV-1 envelope glycoproteins: implications for evasion of immune responses and design of vaccine immunogens. Biochem Biophys Res Commun 390:404–409. https://doi.org/10.1016/j.bbrc.2009.09.029

Yang WP, Green K, Pinz-Sweeney S et al (1995) CDR walking mutagenesis for the affinity maturation of a potent human anti-HIV-1 antibody into the picomolar range. J Mol Biol 254:392–403

Zanetti G, Briggs JAG, Grünewald K et al (2006) Cryo-electron tomographic structure of an immunodeficiency virus envelope complex in situ. PLoS Pathog 2:e83. https://doi.org/10.1371/journal.ppat.0020083

Zebedee SL, Lamb RA (1988) Influenza A virus M2 protein: monoclonal antibody restriction of virus growth and detection of M2 in virions. J Virol 62:2762–2772

Zhang H, Fu H, Luallen RJ et al (2015) Antibodies elicited by yeast glycoproteins recognize HIV-1 virions and potently neutralize virions with high mannose N-glycans. Vaccine 33:5140–5147. https://doi.org/10.1016/j.vaccine.2015.08.012

Zhou T, Georgiev I, Wu X et al (2010) Structural basis for broad and potent neutralization of HIV-1 by antibody VRC01. Science 329:811–817. https://doi.org/10.1126/science.1192819

Zhou T, Lynch RM, Chen L et al (2015) Structural repertoire of HIV-1-neutralizing antibodies targeting the CD4 supersite in 14 donors. Cell 161:1280–1292. https://doi.org/10.1016/j.cell.2015.05.007

Zhou T, Zheng A, Baxa U et al (2018) A neutralizing antibody recognizing primarily N-linked glycan targets the silent face of the HIV envelope. Immunity 48:500–513.e6. https://doi.org/10.1016/j.immuni.2018.02.013

Zhou T, Zhu J, Wu X et al (2013) Multidonor analysis reveals structural elements, genetic determinants, and maturation pathway for HIV-1 neutralization by VRC01-class antibodies. Immunity 39:245–258. https://doi.org/10.1016/j.immuni.2013.04.012

Zhu P, Liu J, Bess J et al (2006) Distribution and three-dimensional structure of AIDS virus envelope spikes. Nature 441:847–852. https://doi.org/10.1038/nature04817

Zou W, Ke J, Zhu J et al (2012) The antigenic property of the H5N1 avian influenza viruses isolated in central China. Virol J 9:148. https://doi.org/10.1186/1743-422X-9-148

Zwick MB, Labrijn AF, Wang M et al (2001) Broadly neutralizing antibodies targeted to the membrane-proximal external region of human immunodeficiency virus type 1 glycoprotein gp41. J Virol 75:10892–10905. https://doi.org/10.1128/JVI.75.22.10892-10905.2001

Immunogenicity and Immunodominance in Antibody Responses

Monique Vogel and Martin F. Bachmann

Contents

Abstract A large number of vaccines exist that control many of the most important infectious diseases. Despite these successes, there remain many pathogens without effective prophylactic vaccines. Notwithstanding strong difference in the biology of these infectious agents, there exist common problems in vaccine design. Many infectious agents have highly variable surface antigens and/or unusually high antibody levels are required for protection. Such high variability may be addressed by using conserved epitopes and these are, however, usually difficult to display with the right conformation in an immunogenic fashion. Exceptionally high antibody titers may be achieved using life vectors or virus-like display of the epitopes. Hence, an important goal in modern vaccinology is to induce high antibody responses against fragile antigens.

M. Vogel · M. F. Bachmann (✉)
Department of Rheumatology, Immunology and Allergology, University Hospital Bern, Sahlihaus 2, CH-3010 Bern, Switzerland
e-mail: Martin.bachmann@me.com

M. Vogel
e-mail: monique.vogel@dbmr.unibe.ch

M. F. Bachmann
The Jenner Institute, University of Oxford, Oxford, UK

Current Topics in Microbiology and Immunology (2020) 428: 89–102
https://doi.org/10.1007/82_2019_160

Published Online: 28 March 2019

1   Introduction

The most successful prophylactic vaccines are based on the induction of neutralizing antibodies. Indeed, almost all currently used vaccines have neutralizing antibodies as their primary protective mechanism. These vaccines profit from two important features of the pathogens they are designed to protect against: (1) low amounts of specific antibodies are usually sufficient for protection and (2) the epitopes recognized by neutralizing antibodies are well exposed and stable. Hence, the prototype antiviral vaccine consists of chemically inactivated viral particles and the prototype antibacterial vaccine is either a chemically inactivated toxin or bacterial carbohydrates coupled to a protein carrier. In many instances, such vaccines induce low levels of specific antibodies, which, however, suffice to protect against infection. Influenza virus may be a point in case, where an antiviral titer of 1:64 (measured by hemagglutination inhibition) is considered to be protective.

The most searched for vaccines are directed against two major types of pathogens, namely, the recently emerging ones and those where the induction of protective antibody levels is difficult (Table 1). The first class has prominent members such as Ebola, MERS, SARS, and emerging influenza strains. The difficulty in making vaccines against these pathogens are not because it would be technically challenging but rather because of the stringent timelines and the often low attack rates, which renders registration studies a difficult task. It is obvious that for a vaccine that usually needs >10 years of development, a virus-like SARS, which caused a dangerous outbreak but disappeared thereafter, is a formidably difficult target. Nevertheless, the recent success with the successful registration of an Ebola vaccine has shown that such problems can be solved in face of an infectious outbreak threatening large parts of the world.

This review will focus on vaccines against pathogens with demanding immunological properties and discuss some of the challenges and possible strategies to cope with them.

Table 1 The two main groups of pathogens with remaining difficulties for development of protective vaccines

	Two types of pathogens difficult to target	
	Complex Biology	Emerging Diseases
Characteristics of antibodies	Generation of neutralizing Ab difficult High amounts of Ab needed Ab may not be protective	Generation of neutralizing Ab simple Low amounts of Ab needed Ab alone protective
Timelines	Not critical	Critical
Examples	Tuberculosis, malaria, HIV	Ebola, MERS, Zika, SARS

2 Optimizing Antibody Responses

2.1 General Considerations

Two important factors have to be taken into consideration for antigen display. Namely, display of native, conformational epitopes with an ability to induce functionally relevant, protective antibodies as well as presenting epitopes in a maximally immunogenic fashion.

Viruses are well known for their ability to induce strong and long-lasting antibody response. The reasons and mechanism causing the strong immunogenicity of viruses have been elucidated in recent years and three key parameters have been identified (Bachmann and Jennings 2010; Jennings and Bachmann 2007): (1) the surface of most viruses is highly repetitive and exhibits an ordered and organized surface structure; (2) viruses are nanoparticles and exhibit the right size to freely reach B cell follicles via lymph in absence of cellular transport for interaction with B cells in a native form; and (3) they are able to trigger toll-like and other innate receptors and activate the complement cascade. This will enhance magnitude and duration of IgG responses as well as cause isotype switching to more protective subclasses.

Thus, viruses and viral immunology may serve as a paradigm for vaccine design and formulation. In general terms, novel vaccines against pathogens with complex immunology should display native epitopes in a highly repetitive fashion on a particulate scaffold, which should be linked to potent adjuvants such as TLR ligands. This will allow induction of immune responses of appropriate magnitude, quality, and specificity (Fig. 1a).

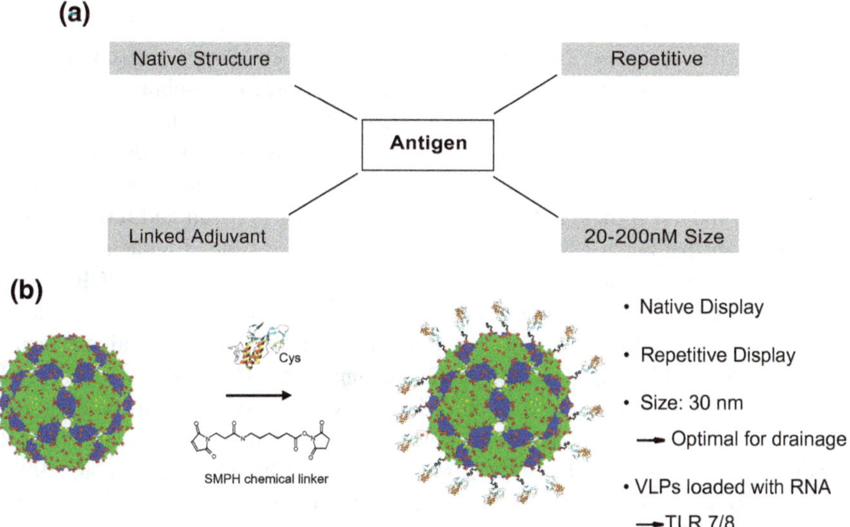

Fig. 1 Four critical parameters for designing vaccines against complex antigens. **a** general outline of the four critical features. **b** outline of one specific solution for vaccine design. Antigens are displayed on RNA-loaded VLPs by means of chemical conjugation

Virus-like particles (VLPs) represent an interesting vaccine design platform that is able to combine many of the elements above into a single system making them extremely potent inducers of neutralizing antibody responses. Indeed, during the last 30 years, vaccines based on VLPs have been the subject of extensive preclinical and clinical research. Several highly efficacious VLP-based vaccines for prevention of infection with hepatitis B or human papillomaviruses are today marketed for human use and a large number of VLP-based vaccines are under development, targeting diseases as different as influenza virus (Lopez-Macias et al. 2011), malaria (Aponte et al. 2007), or Alzheimer`s disease (Winblad et al. 2012). VLPs may be used to induce antibodies against the VLP itself or antigens displayed on them (Fig. 1b).

2.2 Strength of the Responses

Soluble antigens are usually very poorly immunogenic unless formulated in appropriate adjuvants and even then immunogenicity may remain too low for efficacy. It is therefore of interest to display epitopes in an inherently more immunogenic fashion. One possibility is to directly link the antigen to a molecular adjuvant, such as bacterial DNA, an option that will be discussed in the next section. Multimeric display is an additional powerful way to enhance antibody responses. Viral particles usually exhibit highly repetitive and quasi-crystalline surfaces and are known for their ability to induce rapid and strong antibody responses (Bachmann and Zinkernagel 1997). The explanation for the high immunogenicity of repetitive surfaces is as follows. Most viral genomes, in particular RNA viruses, are small and encode relatively few gene products which limit the number of available structural proteins. Therefore, viruses are forced to use multiple copies of a few proteins to assemble their envelopes and cores. As a consequence, viral surfaces consist of ordered and repeated subunits forming a densely packed quasi-crystalline and organized surface. Coevolution of viruses and hosts has resulted in an adaptive immune system rapidly detecting, discriminating, and responding to these repeated and ordered structures found on viral surfaces (Bachmann and Zinkernagel 1996). Hence, antigen organization and repetitiveness is a geometric pathogen-associated molecular pattern. At the cellular level, organized epitopes on the viral surface cross-link specific B cell receptors (BCR) on the surface of B cells. Cross-linking of the BCR resulting in stabilizing BCR-signaling micro-domains constitutes a strong activation signal for B cells and can cause prompt T-independent IgM response (Bachmann et al. 1995; Thyagarajan et al. 2003). Viral proteins expressed in an ordered and repetitive fashion are considerably more immunogenic than in soluble form and can even overcome B cell tolerance (Bachmann et al. 1993; Justewicz et al. 1995; Schodel et al. 1993). The optimal spacing of epitopes for activation of B cells has been analyzed using haptenated polymers and the immunon was defined as 20–25 epitopes spaced by 5–10 nm leading to optimal geometric characteristics for B cell activation (Dintzis et al. 1982). Immunogenic epitopes displayed on viruses and VLPs have similar numbers and distance (Jegerlehner et al. 2002).

There are multiple ways to render antigens highly organized. Some viral proteins spontaneously assemble into VLPs upon expression, constituting a simple way to render the proteins highly repetitive. Indeed, the successful vaccines against hepatitis B and human papillomavirus (HPV) are examples of proteins spontaneously assembling into VLPs. VLPs may be used as scaffolds for antigen displays and genetic fusion of epitopes to the surface of VLPs is an elegant way to give the displayed antigens a "viral fingerprint" by rendering them highly organized (Rhee and Barouch 2009). Although often successful, the genetic fusion approach whereby an antigen is genetically linked to a VLP may limit the size and nature of the epitopes display as complex epitopes often are not compatible with correct assembly of the VLPs. As an alternative, we have developed a method of chemical conjugation, whereby VLPs and antigens are expressed separately and brought together by means of chemical cross-linking. Clinical proof of concept has been achieved for a number of vaccine candidates, including a vaccine against hypertension (Tissot et al. 2008), smoking (Cornuz et al. 2008), allergy (Kundig et al. 2006), and influenza virus (Low et al. 2014). A vaccine against Alzheimer's disease based on a peptide displayed on VLPs is currently undergoing late-stage clinical development (Winblad et al. 2012). An additional method is the SpyCatcher system, whereby autocatalytic peptides are used that bind to each other and undergo covalent linkage (Zakeri et al. 2012; Thrane et al. 2016; Brune et al. 2016). This technology has proven highly efficient at least at lab scale.

There are a number of additional platforms that allow rendering antigens of choice multivalent. Linking antigens to transferrin has been used to render the hemagglutinin of influenza virus highly immunogenic (Santiago et al. 2012). Fusion of antigens to the coiled structure of complement component 3d (C3d) is another way to render antigens that are least pentavalent and to enhance the immune response to various foreign and self antigens (Toapanta and Ross 2006).

Size may also be important for induction of potent B cell responses. Small particles with the dimensions of viruses and VLPs efficiently drain or diffuse to lymph nodes from the site of injection as they can directly enter lymph vessels without the need of cellular transport. Thus, VLPs can enter secondary lymphoid organs and interact directly with B cells to trigger antibody responses. In contrast, larger particles with size >200–500 nm cannot enter the lymphatic system and need to be transported by dendritic cells (Manolova et al. 2008). The fact that active cellular transport by dendritic cells is not required to deliver particles with viral size to B cells means that conformation-dependent epitopes can be presented in a non-processed form and optimal B cell responses may be induced.

2.3 Native Epitope Display

As discussed above, most currently used vaccines are against pathogens for which induction of neutralizing and protective antibodies is relatively simple. As an example, poliovirus displays neutralizing epitopes in a rigid and stable form and attenuated as well as inactivated viruses therefore readily induce neutralizing antibodies. As discussed previously (Bachmann and Zinkernagel 1996), this probably reflects the situation that most viruses, in particular cytolytic viruses, cannot establish a stable long-term relationship with the human host in the absence of rapid induction of neutralizing antibodies protecting the host from lethal infection. This is different for viruses such as HIV, where epitopes with the ability to induce antibodies neutralizing a large number of different viral strains are very unstable and only occur transiently during cellular infection (Karlsson Hedestam et al. 2008). Furthermore, other protective epitopes may be shielded away by, e.g., carbohydrates (e.g., HIV) or protein domains as, e.g., the malaria MSP-1 protein (Guevara Patino et al. 1997)) or the influenza HA stem (Justewicz et al. 1995). Display of such fragile or naturally not exposed epitopes requires careful design going much further than simply using linear peptides as epitope mimic [see, e.g., (Timmerman et al. 2009)]. In addition, testing of large numbers of differently truncated proteins or protein domains may be required to find a version exposing naturally hidden epitopes in a conformationally appropriate way. In both cases, trial and error remains the dominant path forward. Even for a small molecule like nicotine, a large number of different molecules may need to be tested for induction of optimal antibodies (Pryde et al. 2013). In addition, one should keep in mind that binding to a synthetic epitope by a neutralizing antibody does not imply that the same epitope is able to induce such neutralizing antibodies. Specifically, a specific neutralizing antibody may induce a subtle conformational change in the synthetic peptide mimic, which allows optimal binding. Using this same epitope for vaccination may therefore result in a wide range of different antibody specificities with only a small fraction exhibiting the desired neutralizing activity. For this reason, not many vaccine candidates based on such epitopes have successfully entered clinical development.

An additional parameter to consider is the adjuvants used. Alum has, e.g., the tendency to denature fragile antigens resulting in enhanced ELISA titers which not necessarily correlate with enhanced neutralizing antibody responses (Skibinski et al. 2013). We have recently compared Alum to microcrystalline tyrosine (MCT) and found that antibody responses were higher with Alum if assessed by ELISA but the protective capacity of the induced antibodies was higher if MCT was used as the adjuvant (Cabral-Miranda et al. 2017). Native display of antigens on immunogenic carriers, which do not need an adjuvant, may also be an attractive technology to induce potent cellular and humoral responses.

2.4 Duration of the Responses

The innate immune system has been known for a long time to enhance adaptive immunity. Indeed, both B and T cell responses profit from co-engagement of the innate immune system. B cell responses may be influenced directly via B-cell-intrinsic TLR signaling or indirectly via induction of T_{FH} cells by activated dendritic cells (Hou et al. 2011; Jegerlehner et al. 2007). Even Alum, an adjuvant not necessarily associated with specific activation of the innate immune system, has been shown to activate the inflammasome linking Alum to the production of IL-1 (Eisenbarth et al. 2008). The importance of this activation for the adjuvants effect remains, however, controversial (Spreafico et al. 2010). Direct stimulation of B cells by specific delivery of TLR ligands is probably the most potent way to employ the adjuvants. Indeed, conjugation of CpGs to proteins has been shown to be more potent than simple mixing (Bourquin et al. 2008; Eckl-Dorna and Batista 2009; Storni et al. 2004). Furthermore, packaging of RNA or CpGs into viral particles induces potent IgG2a as well as systemic IgA responses in the mouse (Bessa et al. 2008; Jegerlehner et al. 2007). Such particles also induce potent mucosal responses if applied intranasally. Hence, conjugation or packaging of immunostimulatory sequences seems the most potent way to employing them as adjuvants.

Repetitive and organized structures harness the innate immune system in a second way (Bachmann and Jennings 2010). This is due to the fact that repetitive structures are preferentially recognized by components of the complement system and natural antibodies. This results in enhanced phagocytosis and antigen processing, increasing induction of T cell help (Fanger et al. 1996; Matsushita and Fujita 2001). As induction of GC reactions, memory B cells and long-lived plasma cells are Th cell dependent, and this may result in overall increased antibody responses (Bachmann and Zinkernagel 1997). An additional and equally important consequence of binding natural IgM antibodies and complement has enhanced deposition of repetitive structures such as viral particles on FDCs in germinal centers. Indeed, B cells have been shown to bind viral particles to their surface complement receptors CD21 and transport them to FDCs within B cell follicles, driving enhanced GC reactions (Link et al. 2012; Phan et al. 2007). Repetitive antigens may also be transported bound to low-affinity BCRs on B cells for deposition on FDCs (Bessa et al. 2012). In addition, CD21 engagement by complement components bound to repetitive surfaces enhances expression of the plasma cell-specific transcription factors blimp and XBP-1, enabling generation of long-lived plasma cells and prolonging IgG responses (Gatto et al. 2005). Furthermore, complement bound to VLPs reduces the threshold required for B cell activation (Jegerlehner et al. 2002).

3 A Case Study: Influenza Virus

Influenza viruses are RNA viruses and are divided into three different types, namely, influenza A, B, and C viruses. Types B and C are almost exclusively found in humans, whereas influenza A virus circulates in a large variety of warm-blooded animals including birds, pigs, and humans. Influenza A virus is further classified by the antigenic characterization of two major surface glycoproteins, hemagglutinin (HA), and neuraminidase (NA) (Krammer et al. 2015). HA is a lectin responsible for the binding of the virus to target cells via sialic acids, whereas NA is involved in the delivery of the virus from the cells by cleaving sialic acid from the infected cells to allow release of infectious particles. These two proteins are used to classify influenza virus into different subtypes. Up to now 18 different HA and 9 different NA have been identified and the first three HA, H1, H2, H3 are commonly found in humans. HA is a homotrimer consisting of a globular head, HA1 that contains the major antigenic sites, and a stalk/stem region, HA2, that anchors the protein in the virus membrane and contains the viral fusion machinery. While HA1 is highly variable, HA2 is very well conserved. Since HA is the protein involved in the entry process, it is also the primary target for neutralizing antibodies elicited by conventional influenza vaccines. Most of these HA-specific neutralizing antibodies are, however, strain specific because they bind to HA1 whose antigenic sites are subjected to antigenic drift in circulating strains. This antigenic drift makes it necessary to generate new vaccines almost every year and the duration of protection is short. Furthermore, these vaccines do not protect against novel pandemic strains as once a novel pandemic virus is identified it takes several months to develop the new vaccine and some more weeks are required for induction of protective immune responses. Induction of long-term protective immunity not affected by antigenic drift is therefore a major goal in vaccine development. Displaying HA1 on VLPs has proven successful to induce protective antibody responses in mice (Jegerlehner et al. 2013) and humans (Low et al. 2014). This type of vaccine, however, suffers from the same susceptibility to antigenic drift as the classical vaccines. A vaccine targeting either the stalk domain (HA2) or the extracellular domain of the M2 ion channel protein, which is also highly conserved, would be promising targets to develop a universal influenza vaccine and overcome the conserved drift of seasonal influenza virus vaccines. Recent studies have demonstrated the feasibility of using stalk domain-based vaccines. Antibodies against stalk domain have been shown to bind to either group 1 or group 2 hemagglutinins (Sui et al. 2009), both (Corti et al. (2011; Ekiert et al. 2009; Wyrzucki et al. 2015), or even viruses for both the A and B genera (Dreyfus et al. 2012). Using recombinant DNA technology, HA constructs devoid of the highly immunogenic head domain were expressed on the cell surfaces. Co-expression of an HIV Gag-based construct with headless HA constructs lead to the production of chimeric headless HLA virus-like particles (VLPs) (Smith et al. 2013). Immunization of mice with these headless HA provided protection with moderate body weight loss against a lethal homologous viral challenge. Moreover, HA headless VLP immune sera showed reactivity to heterologous strains suggesting that vaccination with headless HA might induce protection against different influenza strains. Another study using chimeric HA constructs showed that by repeated immunization of mice, a broad protection against a variety of influenza virus strains could be elicited. However, up to now no vaccines have been developed that are able to induce protective anti-stalk antibodies regardless of HA subtypes. The extracellular domain of M2 is a poorly immunogenic and different approaches were tested to link M2 to carrier molecules or/and to use potent adjuvants to increase immunogenicity (Kang et al. 2012; Song et al. 2011). In particular, a virus-like particle (VLP)-based vaccine targeting M2 has been described to protect mice from lethal influenza infection (Jegerlehner et al.

2004). Further studies analyzing M2-specfic antibodies showed that in contrast to conventional neutralizing antiviral antibodies, the M2 antibodies rely on their Fc part that binds to Fc receptor and to complement to be protective. Induction of potent IgG subclasses, such as IgG2a in the mouse, is therefore critical for this type of vaccine (Schmitz et al. 2012). Further studies investigating the protective potential of M2-specific antibodies in humans are, however, needed to evaluate the potential efficacy of this candidate antigen as universal vaccine.

4 A Case Study: HIV

The trimeric envelope glycoprotein of HIV is a complex protein that is highly glycosylated. The protein is cleaved from a 160 kD precursor protein into gp41 and gp120 which exists as a trimer. This trimer is, however, rather unstable if expressed recombinantly. Furthermore, the trimeric form of gp120/gp41 can adopt different conformational states (conformers), which may induce different levels of neutralizing antibodies. Chemical stabilization of individual conformers may allow focusing of the response to broadly neutralizing antibodies (Schiffner et al. 2016).

There are three types of gp120-specific antibodies, namely, binding but not neutralizing antibodies, strain-specific neutralizing antibodies, and broadly neutralizing antibodies. Non-neutralizing antibodies recognize epitopes primarily exposed on non-native gp120. This class of antibodies is very abundant, as the gp120 trimer is inherently unstable. The strain-specific antibodies recognize epitopes on the native trimer (Burton and Mascola 2015), and these epitopes are, however, very variable and show great sequence variability. As these epitopes are usually well exposed, strain-specific neutralizing antibodies are also abundant. The broadly neutralizing antibodies also recognize epitopes on the native trimer (Burton and Mascola 2015), which are usually buried within a sea of variability and often not well accessible. The epitopes recognized by these most potent antibodies may be categorized into four classes: the CD4-binding site, two additional sites on the gp120, and a single site on gp41 (Sattentau 2014). It is the goal of all HIV vaccine strategies to induce such broadly neutralizing antibodies.

These features of HIV-GP pose a whole array of problems. First, the carbohydrates shield the protein effectively from the induction of neutralizing antibodies and serve as "dummy" antigens, distracting the immune system and preventing induction of more protective antibodies. Removal of glycosylation sites on proteins designed for vaccination may overcome this problem at least in part. Cleavage of the mature GP160 into GP41 and GP120 results in an unstable protein which readily falls apart. It has recently been shown that the complex can be stabilized by linking the gp41 and gp120 proteins by a disulphide bond (Sanders et al. 2013), resulting in the induction of more broadly neutralizing antibodies upon immunization with these modified trimers. The most difficult feature of the glycoprotein is the various conformational states the trimer may adopt. Attempts to stabilize the closed structure using chemical cross-linkers look promising but still does not easily allow induction of broadly neutralizing antibodies (Schiffner et al. 2016). An additional hurdle for induction of strong neutralizing antibody responses is the similarity of two important neutralizing epitopes with self-molecules, which renders induction of antibody responses difficult due to B cell tolerance (Yang et al. 2013).

Display of functional gp41/gp120 spikes displaying broadly neutralizing epitopes in an immunogenic manner is therefore the goal of current vaccine strategies. Display of separately expressed and stabilized trimers on VLPs by means of, e.g., chemical conjugation may be one way to achieve this goal. As highly repetitive arrays of antigens may overcome B cell tolerance (Bachmann et al. 1993), and this may additionally solve the issue with B cell repertoire tolerized by the cross-reactive self-antigens (Schiller and Chackerian 2014). Alternatively, the stabilized spikes may be formulated in an adjuvant that enhances immune responses without impairing the structure of the fragile epitopes.

Hence, designing an immunogen that displays epitopes of HIV-GP that allow induction of broadly neutralizing antibodies remains a formidable task but appears more feasible now than a few years ago.

References

Aponte JJ, Aide P, Renom M, Mandomando I, Bassat Q, Sacarlal J, Manaca MN, Lafuente S, Barbosa A, Leach A, Lievens M, Vekemans J, Sigauque B, Dubois MC, Demoitie MA, Sillman M, Savarese B, Mcneil JG, Macete E, Ballou WR, Cohen J, Alonso PL (2007) Safety of the RTS, S/AS02D candidate malaria vaccine in infants living in a highly endemic area of Mozambique: a double blind randomised controlled phase I/IIb trial. Lancet 370(9598):1543–1551

Bachmann MF, Hengartner H, Zinkernagel RM (1995) T helper cell-independent neutralizing B cell response against vesicular stomatitis virus: role of antigen patterns in B cell induction? Eur J Immunol 25(12):3445–3451

Bachmann MF, Jennings GT (2010) Vaccine delivery: a matter of size, geometry, kinetics and molecular patterns. Nat Rev Immunol 10(11):787–796

Bachmann MF, Rohrer UH, Kundig TM, Burki K, Hengartner H, Zinkernagel RM (1993) The influence of antigen organization on B cell responsiveness. Science 262(5138):1448–1451

Bachmann MF, Zinkernagel RM (1996) The influence of virus structure on antibody responses and virus serotype formation. Immunol Today 17(12):553–558

Bachmann MF, Zinkernagel RM (1997) Neutralizing antiviral B cell responses. Annu Rev Immunol 15:235–270

Bessa J, Schmitz N, Hinton HJ, Schwarz K, Jegerlehner A, Bachmann MF (2008) Efficient induction of mucosal and systemic immune responses by virus-like particles administered intranasally: implications for vaccine design. Eur J Immunol 38(1):114–126

Bessa J, Zabel F, Link A, Jegerlehner A, Hinton HJ, Schmitz N, Bauer M, Kündig TM, Saudan P, Bachmann MF (2012) Low-affinity B cells transport viral particles from the lung to the spleen to initiate antibody responses. Proc Natl Acad Sci U S A. 109(50):20566–20571

Bourquin C, Anz D, Zwiorek K, Lanz AL, Fuchs S, Weigel S, Wurzenberger C, Von Der Borch P, Golic M, Moder S, Winter G, Coester C, Endres S (2008) Targeting CpG oligonucleotides to the lymph node by nanoparticles elicits efficient antitumoral immunity. J Immunol 181 (5):2990–2998

Brune KD, Leneghan DB, Brian IJ, Ishizuka AS, Bachmann MF, Draper SJ, Biswas S, Howarth M (2016) Sci Rep 19(6):19234

Burton DR, Mascola JR (2015) Antibody responses to envelope glycoproteins in HIV-1 infection. Nat Immunol 16(6):571–576

Cabral-Miranda G, Heath MD, Gomes AC, Mohsen MO, Montoya-Diaz E, Salman AM, Atcheson E, Skinner MA, Kramer MF, Reyes-Sandoval A, Bachmann M (2017) Microcrystalline Tyrosine (MCT®): a depot adjuvant in licensed allergy immunotherapy offers new opportunities in malaria. Vaccine 5(4)

Cornuz J, Zwahlen S, Jungi WF, Osterwalder J, Klingler K, Van Melle G, Bangala Y, Guessous I, Muller P, Willers J, Maurer P, Bachmann MF, Cerny T (2008) A vaccine against nicotine for smoking cessation: a randomized controlled trial. PLoS ONE 3(6):e2547

Corti D, Voss J, Gamblin SJ, Codoni G, Macagno A et al (2011) A neutralizing antibody selected from plasma cells that binds to group 1 and group 2 influenza a hemagglutinins. Science 333 (6044):850–56. https://doi.org/10.1126/science.1205669

Dintzis RZ, Vogelstein B, Dintzis HM (1982) Specific cellular stimulation in the primary immune response: experimental test of a quantized model. Proc Natl Acad Sci U S A 79(3):884–888

Dreyfus C, Laursen NS, Kwaks T, Zuijdgeest D, Khayat R et al (2012) Highly conserved protective epitopes on influenza B viruses. Science 337(6100):1343–48. https://doi.org/10.1126/science.1222908

Eckl-Dorna J, Batista FD (2009) BCR-mediated uptake of antigen linked to TLR9 ligand stimulates B-cell proliferation and antigen-specific plasma cell formation. Blood 113 (17):3969–3977

Eisenbarth SC, Colegio OR, O'connor W, Sutterwala FS, Flavell RA (2008) Crucial role for the Nalp3 inflammasome in the immunostimulatory properties of aluminium adjuvants. Nature 453 (7198):1122–6

Ekiert, DC, Bhabha G, Elsliger MA, Friesen RHE, Jongeneelen M, others (2009) Antibody Recognition of a Highly Conserved Influenza Virus Epitope. Science, 324.5924:246–51. http://dx.doi.org/10.1126/science.1171491

Fanger NA, Wardwell K, Shen L, Tedder TF, Guyre PM (1996) Type I (CD64) and type II (CD32) Fc gamma receptor-mediated phagocytosis by human blood dendritic cells. J Immunol 157 (2):541–548

Gatto D, Pfister T, Jegerlehner A, Martin SW, Kopf M, Bachmann MF (2005) Complement receptors regulate differentiation of bone marrow plasma cell precursors expressing transcription factors Blimp-1 and XBP-1. J Exp Med 201(6):993–1005

Guevara Patino JA, Holder AA, Mcbride JS, Blackman MJ (1997) Antibodies that inhibit malaria merozoite surface protein-1 processing and erythrocyte invasion are blocked by naturally acquired human antibodies. J Exp Med 186(10):1689–1699

Hou B, Saudan P, Ott G, Wheeler ML, Ji M, Kuzmich L, Lee LM, Coffman RL, Bachmann MF, Defranco AL (2011) Selective utilization of Toll-like receptor and MyD88 signaling in B cells for enhancement of the antiviral germinal center response. Immunity 34(3):375–384

Jegerlehner A, Maurer P, Bessa J, Hinton HJ, Kopf M, Bachmann MF (2007) TLR9 signaling in B cells determines class switch recombination to IgG2a. J Immunol 178(4):2415–2420

Jegerlehner A, Schmitz N, Storni T, Bachmann MF (2004) Influenza A vaccine based on the extracellular domain of M2: weak protection mediated via antibody-dependent NK cell activity. J Immunol 172(9):5598–5605

Jegerlehner A, Storni T, Lipowsky G, Schmid M, Pumpens P, Bachmann MF (2002) Regulation of IgG antibody responses by epitope density and CD21-mediated costimulation. Eur J Immunol 32(11):3305–3314

Jegerlehner A, Zabel F, Langer A, Dietmeier K, Jennings GT, Saudan P, Bachmann MF (2013) Bacterially produced recombinant influenza vaccines based on virus-like particles. PLoS ONE 8(11):e78947

Jennings GT, Bachmann MF (2007) Designing recombinant vaccines with viral properties: a rational approach to more effective vaccines. Curr Mol Med 7(2):143–155

Justewicz DM, Doherty PC, Webster RG (1995) The B-cell response in lymphoid tissue of mice immunized with various antigenic forms of the influenza virus hemagglutinin. J Virol 69 (9):5414–5421

Kang SM, Kim MC, Compans RW (2012) Virus-like particles as universal influenza vaccines. Expert Rev Vaccines 11(8):995–1007

Karlsson Hedestam GB, Fouchier RA, Phogat S, Burton DR, Sodroski J, Wyatt RT (2008) The challenges of eliciting neutralizing antibodies to HIV-1 and to influenza virus. Nat Rev Microbiol 6(2):143–155

Krammer F, Palese P, Steel J (2015) Advances in universal influenza virus vaccine design and antibody mediated therapies based on conserved regions of the hemagglutinin. Curr Top Microbiol Immunol 386:301–321

Kundig TM, Senti G, Schnetzler G, Wolf C, Prinz Vavricka BM, Fulurija A, Hennecke F, Sladko K, Jennings GT, Bachmann MF (2006) Der p 1 peptide on virus-like particles is safe and highly immunogenic in healthy adults. J Allergy Clin Immunol 117(6):1470–1476

Link A, Zabel F, Schnetzler Y, Titz A, Brombacher F, Bachmann MF (2012) Innate immunity mediates follicular transport of particulate but not soluble protein antigen. J Immunol 188 (8):3724–3733

Lopez-Macias C, Ferat-Osorio E, Tenorio-Calvo A, Isibasi A, Talavera J, Arteaga-Ruiz O, Arriaga-Pizano L, Hickman SP, Allende M, Lenhard K, Pincus S, Connolly K, Raghunandan R, Smith G, Glenn G (2011) Safety and immunogenicity of a virus-like particle pandemic influenza A (H1N1) 2009 vaccine in a blinded, randomized, placebo-controlled trial of adults in Mexico. Vaccine 29(44):7826–7834

Low JG, Lee LS, Ooi EE, Ethirajulu K, Yeo P, Matter A, Connolly JE, Skibinski DA, Saudan P, Bachmann M, Hanson BJ, Lu Q, Maurer-Stroh S, Lim S, Novotny-Diermayr V (2014) Safety and immunogenicity of a virus-like particle pandemic influenza A (H1N1) 2009 vaccine: results from a double-blinded, randomized Phase I clinical trial in healthy Asian volunteers. Vaccine 32(39):5041–5048

Manolova V, Flace A, Bauer M, Schwarz K, Saudan P, Bachmann MF (2008) Nanoparticles target distinct dendritic cell populations according to their size. Eur J Immunol 38(5):1404–1413

Matsushita M, Fujita T (2001) Ficolins and the lectin complement pathway. Immunol Rev 180:78–85

Phan TG, Grigorova I, Okada T, Cyster JG (2007) Subcapsular encounter and complement-dependent transport of immune complexes by lymph node B cells. Nat Immunol 8(9): 992–1000

Pryde DC, Jones LH, Gervais DP, Stead DR, Blakemore DC, Selby MD, Brown AD, Coe JW, Badland M, Beal DM, Glen R, Wharton Y, Miller GJ, White P, Zhang N, Benoit M, Robertson K, Merson JR, Davis HL, Mccluskie MJ (2013) Selection of a novel anti-nicotine vaccine: influence of antigen design on antibody function in mice. PLoS ONE 8(10):e76557

Rhee EG, Barouch DH (2009) Translational mini-review series on vaccines for HIV: harnessing innate immunity for HIV vaccine development. Clin Exp Immunol 157(2):174–180

Sanders RW, Derking R, Cupo A, Julien J, Yasmeen A, De Val N, Kim HJ, Blattner C, De La Pena AT, Korzun J, Gobalek M, De Los Reyes K, Ketas T, Van Gils M, King C, Wilson I, Ward AB, Klasse P, Moore JP (2013) A next generation cleaved, soluble HIV-1 Env trimer, BG505 SOSIp.664 gp140, expresses multiple epitopes for broadly neutralizing but non-neutralizing antibodies. PLos Pathog 9(9)(e1003618)

Santiago FW, Lambert Emo K, Fitzgerald T, Treanor JJ, Topham DJ (2012) Antigenic and immunogenic properties of recombinant hemagglutinin proteins from H1N1 A/Brisbane/59/07 and B/Florida/04/06 when produced in various protein expression systems. Vaccine 30 (31):4606–4616

Sattentau QJ (2014) Immunogen design to focus the B-cell repertoire. Curr Opin HIV AIDS 9 (3):217–223

Schiffner T, De Val N, Russell RA, De Taeye SW, De La Pena AT, Ozorowski G, Kim HJ, Nieusma T, Brod F, Cupo A, Sanders RW, Moore JP, Ward AB, Sattentau QJ (2016) Chemical cross-linking stabilizes native-like HIV-1 envelope glycoprotein trimer antigens. J Virol 90 (2):813–828

Schiller J, Chackerian B (2014) Why HIV virions have low numbers of envelope spikes: implications for vaccine development. PLos Pathog 10(8)(e1004254)

Schmitz N, Beerli RR, Bauer M, Jegerlehner A, Dietmeier K, Maudrich M, Pumpens P, Saudan P, Bachmann MF (2012) Universal vaccine against influenza virus: linking TLR signaling to anti-viral protection. Eur J Immunol 42(4):863–869

Schodel F, Peterson D, Zheng J, Jones JE, Hughes JL, Milich DR (1993) Structure of hepatitis B virus core and e-antigen. A single precore amino acid prevents nucleocapsid assembly. J Biol Chem 268(2):1332–7

Skibinski DA, Hanson BJ, Lin Y, Von Messling V, Jegerlehner A, Tee JB, De Chye H, Wong SK, Ng AA, Lee HY, Au B, Lee BT, Santoso L, Poidinger M, Fairhurst AM, Matter A, Bachmann MF, Saudan P, Connolly JE (2013) Enhanced neutralizing antibody titers and Th1 polarization from a novel Escherichia coli derived pandemic influenza vaccine. PLoS ONE 8 (10):e76571

Smith GE, Flyer DC, Raghunandan R, Liu Y, Wei Z, Wu Y, Kpamegan E, Courbron D, Fries LF 3rd, Glenn GM (2013) Development of influenza H7N9 virus like particle (VLP) vaccine: homologous A/Anhui/1/2013 (H7N9) protection and heterologous A/chicken/Jalisco/CPA1/ 2012 (H7N3) cross-protection in vaccinated mice challenged with H7N9 virus. Vaccine 31 (40):4305–4313

Song JM, Van Rooijen N, Bozja J, Compans RW, Kang SM (2011) Vaccination inducing broad and improved cross protection against multiple subtypes of influenza A virus. Proc Natl Acad Sci U S A 108(2):757–761

Spreafico R, Ricciardi-Castagnoli P, Mortellaro A (2010) The controversial relationship between NLRP3, alum, danger signals and the next-generation adjuvants. Eur J Immunol 40(3):638–642

Storni T, Ruedl C, Schwarz K, Schwendener RA, Renner WA, Bachmann MF (2004) Nonmethylated CG motifs packaged into virus-like particles induce protective cytotoxic T cell responses in the absence of systemic side effects. J Immunol 172(3):1777–1785

Sui J, Hwang WC, Perez S, Wei G, Aird D, Chen LM, Santelli E, Stec B, Cadwell G, Ali M, Wan H, Murakami A, Yammanuru A, Han T, Cox NJ, Bankston LA, Donis RO, Liddington RC, Marasco WA (2009) Structural and functional bases for broad-spectrum neutralization of avian and human influenza A viruses. Nat Struct Mol Biol 16(3):265–273

Thrane S, Janitzek CM, Matondo S, Resende M, GustavssonT, de Jongh WA, Clemmensen S, Roeffen W, van de Vegte Bolmer M, van Gemert GJ, Sauerwein R, Schiller JT, NielsenMA, Theander TG, Salanti A, SanderAF (2016) J Nanobiotechnol (2016)14:30

Thyagarajan R, Arunkumar N, Song W (2003) Polyvalent antigens stabilize B cell antigen receptor surface signaling microdomains. J Immunol 170(12):6099–6106

Timmerman P, Puijk WC, Boshuizen RS, Van Dijken P, Slootstra JW, Beurskens FJ, Parren PWHI, Huber A, Bachmann MF, Meloen RH (2009) Functional Reconstruction of structurally complex epitopes using CLIPSTM technology. The Open Vaccine J 2:56–67

Tissot AC, Maurer P, Nussberger J, Sabat R, Pfister T, Ignatenko S, Volk HD, Stocker H, Muller P, Jennings GT, Wagner F, Bachmann MF (2008) Effect of immunisation against angiotensin II with CYT006-AngQb on ambulatory blood pressure: a double-blind, randomised, placebo-controlled phase IIa study. Lancet 371(9615):821–827

Toapanta FR, Ross TM (2006) Complement-mediated activation of the adaptive immune responses: role of C3d in linking the innate and adaptive immunity. Immunol Res 36(1–3):197–210

Winblad B, Andreasen N, Minthon L, Floesser A, Imbert G, Dumortier T, Maguire RP, Blennow K, Lundmark J, Staufenbiel M, Orgogozo JM, Graf A (2012) Safety, tolerability, and antibody response of active Abeta immunotherapy with CAD106 in patients with Alzheimer's disease: randomised, double-blind, placebo-controlled, first-in-human study. Lancet Neurol 11 (7):597–604

Wyrzucki A, Bianchi M, Kohler I, Steck M, Hangartner L (2015) Heterosubtypic antibodies to influenza a virus have limited activity against cell-bound virus but are not impaired by strain-specific serum antibodies. J Virol 89.6 [Sandri-Goldin M (ed) American Society for Microbiology 3136–44] http://dx.doi.org/10.1128/JVI.03069-14

Yang G, Holl TM, Liu Y, Li Y, Lu X, Nicely NI, Kepler TB, Alam SM, Liao HX, Cain DW, Spicer L, Vandeberg JL, Haynes BF, Kelsoe G (2013) Identification of autoantigens recognized by the 2F5 and 4E10 broadly neutralizing HIV-1 antibodies. J Exp Med 210 (2):241–256
Zakeri B, Fierer JO, Celik E, Chittock EC, Schwarz-Linek U, Moy VT, Howarth M (2012) Peptide tag forming a rapid covalent bond to a protein, through engineering a bacterial adhesin. Proc Natl Acad Sci U S A 109(12):E690–E697

Adjuvants

Darrick Carter, Malcolm S. Duthie and Steven G. Reed

Contents

Abstract Developing new vaccines against emerging pathogens or pathogens where variability of antigenic sites presents a challenge, the inclusion of stimulators of the innate immune system is critical to mature the immune response in a way that allows high avidity recognition while preserving the ability to react to drifted serovars. The innate immune system is an ancient mechanism for recognition of nonself and the first line of defense against pathogen insult. By triggering innate receptors, adjuvants can boost responses to vaccines and enhance the quality and magnitude of the resulting immune response. This chapter: (1) describes the innate

D. Carter (✉)
PAI Life Sciences Inc., 1616 Eastlake Ave E, Suite 550, Seattle,
WA 98102, USA
e-mail: Darrick.Carter@idri.org

D. Carter · M. S. Duthie · S. G. Reed
Adjuvant Technologies, IDRI, 1616 Eastlake Avenue E., Suite 400, Seattle,
WA 98102, USA

D. Carter · M. S. Duthie · S. G. Reed
Global Health, University of Washington, 1616 Eastlake Ave E, Suite 400,
Seattle, WA 98102, USA

Current Topics in Microbiology and Immunology (2020) 428: 103–127
https://doi.org/10.1007/82_2018_112
© Springer International Publishing AG, part of Springer Nature 2018
Published Online: 25 July 2018

immune system, (2) provides examples of how adjuvants are formulated to optimize their effectiveness, and (3) presents examples of how adjuvants can improve outcomes of immunization.

1 The Innate Immune System

There are two widely recognized phases of the immune response to invading pathogens: the innate response that protects the host rapidly and nonspecifically; and an adaptive response, where the host specifically targets antigens encoded in the pathogen to mount a long lasting, protective response. Innate immunity is conferred primarily through Pattern Recognition Receptors (PRRs) that are triggered by Microbe-Associated Molecular Patterns (MAMPs) (Degn and Thiel 2013; Ray et al. 2013). This system effectively combines two disparate needs for a receptor–ligand interaction: it has broad enough recognition to respond to an array of diverse patterns found across a wide variety of invading organisms, but specific enough to ignore similar molecules found within the host. Further elegance of this system is that the molecular patterns that are recognized are generally indispensable for the survival of the invading organism.

Triggering the innate system leads to rapid release of a finely tuned composition of cytokines and chemokines that allows the appropriate effector cells to attempt to quell the infecting microbe or to alter the immune response to allergens. The nature of the antibody response to allergens can lead to immunity or allergy. While low dose antigen is known to induce Th1 responses (Bretscher et al. 1992), including the TLR4 ligand MonoPhosphoryl Lipid A, "MPL" with allergen can similarly skew the antibody response qualitatively, leading to the approval of MPL-adjuvanted allergens for human use (Puggioni et al. 2005; McCormack and Wagstaff 2006; Gawchik and Saccar 2009; Rosewich et al. 2013). As the human distribution of TLR4 is restricted to antigen-presenting cells, it is evident that early innate signaling influences the adaptive response. These educational interactions through which cells from the adaptive immune system are recruited and matured through presentation of antigens in the context of proper signaling molecules occur predominantly in lymph nodes proximate to, and draining, the site of the initial innate interactions. In studies, where antigens are presented to the host—with or without adjuvants that provide the innate stimulation—the importance of concomitant innate signaling for expansion of cell populations, effective formation of germinal centers (GCs), and accelerated hypermutation of B cells is becoming clear (Capolunghi et al. 2013; Carter and Reed 2010; Palm and Medzhitov 2009).

2 Pattern Recognition Receptors

The innate immune system has pattern recognition mechanisms that allow the rapid sensing of danger signals from pathogens. These Pathogen Recognition Receptors "PRRs" rely on widely distributed, but conserved, molecular families found on pathogens (called Pathogen Associated Molecular Patterns, "PAMPs") (Foldes et al. 2010). Activation of PRR by their cognate ligands leads to production of inflammatory cytokines, upregulation of MHC and costimulatory molecules on APC, activation of NK cells, and priming and amplification of antigen-specific T- and B-cell responses. The Toll-Like Receptor (TLR) family members are probably the most well-known PRR, although further non-TLR PRR, such as the Nucleotide-Binding Oligomerization Domain (NOD), leucine-rich repeat-containing family (or NOD-Like Receptors, "NLR") and the retinoic acid-inducible protein 1 (RIG-I-Like Receptors; RLR), are now being studied (Martinon and Tschopp 2005; Kang et al. 2002). As NLR and RLR ligands are recognized, and the interplay between these and TLR ligands can be examined, it is likely that novel adjuvant formulations will be generated to capitalize on synergistic activities between these molecules.

At present, the TLR family is certainly the most widely studied and defined PRR, with TLR ligands being developed through to licensed products. The transmembrane-located toll receptor was first identified in *Drosophila*, where its function is required for protection against fungal and Gram-positive bacterial infections (Imler and Zheng 2004). Several TLR, and the Toll/Interleukin (IL)-1 Receptor (TIR) domain that is a shared signaling motif of the IL-1 receptor and TLR have subsequently been identified in mammals (O'Neill and Bowie 2007). The TLR family currently consists of 13 individual molecules, 10 of which are found in humans. Equivalents of certain TLR found in humans are not present in all mammals, complicating or confounding the use of experimental animals to predict human responses to TLR agonists. Binding of TLR/TIR leads to recruitment of signaling adaptors to the TLR/TIR interface: MyD88, MyD88 adaptor-like (MAL/TIRAP), TIR domain-containing adaptor protein-inducing interferon (IFN)-β (TRIF, which is also known as TICAM1), TRIF-related adaptor molecule (TRAM, which is also known as TICAM2), and sterile α-and armadillo motif-containing protein (SARM). Signaling through MyD88 can induce type I IFN (IFN-α and IFN-β) and pro-inflammatory cytokines such as Tumor Necrosis Factor (TNF) and IL-6 (O'Neill and Bowie 2007) and in the absence of MyD88 signaling through TLR2, TLR5, TLR7, TLR8, and TLR9 is lost. In contrast, TLR3 signaling is mediated exclusively by TRIF (O'Neill and Bowie 2007), while TLR4 is unique in using both the MyD88 and TRIF signaling pathways. Like MyD88, signaling through TRIF induces activation of NFκB and expression of IFN-β.

TLR form dimers to allow binding of, and triggering by, a diverse array of molecules. TLR1 and TLR6 heterodimerize with TLR2 to recognize di- and tri-acylated lipopeptides, as well as certain acylated sugars and bacterial polysaccharides (Basith et al. 2011; Thoma-Uszynski et al. 2001). TLR3, TLR7, and TLR8 recognize specific forms of RNA (Basith et al. 2011; Breckpot et al. 2010; Zucchini et al. 2008) while TLR9 recognizes CpG motifs in DNA (Zucchini et al. 2008; Hemmi et al. 2000; Hemmi et al. 2003; Johnson et al. 2009). Finally, TLR5 identifies bacterial flagellins (Foldes et al. 2010; Hayashi et al. 2001) while TLR4 can bind to lipopolysaccharides (LPSs) and lipid A family members (Anderson et al. 2010; Persing et al. 2002; Baldridge et al. 2004; Mata-Haro et al. 2007; El Shikh et al. 2007).

Signaling through TLR is sufficient for the initiation of adaptive immune responses by Antigen-Presenting Cells (APC) (Palm and Medzhitov 2009), but both the cellular location of TLR and the cell type expressing TLR could strongly influence the adjuvant formulations to include within certain vaccines. TLR family members are differentially expressed by a wide variety of hematopoietic and non-hematopoietic cell types, and can be located on the cell surface and within endosomal compartments (Palm and Medzhitov 2009; Gribar et al. 2008; Chang 2010). While TLR function is best described in dendritic cells and macrophages, signaling directly through TLR expressed on B- or T cells can support antibody class switching, activation of memory B cells, suppression of the T regulatory cell response, and the survival of effector T cells.

TLR1, TLR2, and TLR6

TLR2 forms heterodimers with either TLR1 or TLR6 in mice and with TLR1, TLR6, or TLR10 in humans (Ozinsky et al. 2000; Takeuchi et al. 2001; Hasan et al. 2005). There are numerous known ligands for TLR2 including peptidoglycans from the cell walls of Gram-Positive bacteria (PGN), bacterial lipopeptides and glycolipids, lipoarabinomannans from *Mycobacteria* and parasitic GPI-anchors. TLR2-targeted adjuvants include agonists like the tri-acylated lipopeptides (e.g., Pam3CysSerLys4 and derivatives), as well as Group B *Streptococcus*-derived polysaccharides which have been used in several experimental models including adjuvanting tetanus toxoid. Using these TLR2/TLR1 agonist-based adjuvants can support T-cell priming, enhance B-cell responses, and results in robust tetanus-specific antibody titers (Gallorini et al. 2009). Similarly, formulations of the TLR2/TLR6-targeted macrophage-activating lipopeptide (MALP)–2 stimulated strong antibody and cellular responses, while intranasal administration of a synthetic MALP-2 in combination with the *E. coli*-derived antigens EspB and C-280 γ-Intimin elicited both serum and mucosal antibodies (Cataldi et al. 2008), other reports indicated that MALP-2 analogs could be combined with antigens to elicit cellular responses characterized by specific secretion of the cytokines IFN-γ, IL-4 and IL-2 (Prajeeth et al. 2010). In cultured human lung tissue, S-[2,3-bispalmitoyiloxy-(2R)-propyl]-R-cystenyl-amido-monomethoxyl polyethylene glycol ("BPP")—another synthetic analog of MALP-2-targeting TLR2—had activity at inducing TNF, IL-10, and MIP-1β and in mice. this molecule was shown to increase cross-presentation by dendritic cells and thereby, increase CD8-mediated cytotoxic T-cell responses (Phillipps et al. 2009). These studies demonstrate that TLR2-directed adjuvants have the potential to increase responses to co-formulated antigens and although TLR2 agonist-based adjuvants have not been approved in licensed products for administration in humans, they have the potential for further translation through the clinic. This is, especially, true as many can be directly conjugated to antigens, a promising direction for new adjuvant administration strategies such as the lipid core peptide that mimics Pam3CSK, and which is being developed to allow the simultaneous production of the adjuvant with the antigen (Phillipps et al. 2009).

TLR3

TLR3 recognizes Double-Stranded Ribonucleic Acids (dsRNAs) and is localized to the interior of endosomes. Its putative function is to detect these nucleic acids when produced by replicating viruses in an infected cell and alert the immune system to the presence of this viral threat. poly(I:C) is a synthetic polymer that mimics the viral dsRNAs and binds TLR3 to induce the production of IFN-β (Alexopoulou et al. 2001). TLR3 signals exclusively through TRIF/TRAM leading to activation of NF-κB regulated gene expression and resulting in a Th1-polarized immune response that is protective against viral infection. TLR3 ligation may also be required for activation of RIG-I (Palm and Medzhitov 2009), providing complementary pathways for detecting and responding to viral products within the cell. Due to their ability to induce interferon responses, TLR3 agonists are particularly appealing as adjuvant candidates for immune strategies targeting intracellular pathogens and cancers (O'Neill et al. 2009).

TLR4

TLR4 expression is found in a variety of cell types in most animals (Gribar et al. 2008; Dasari et al. 2005). Unlike in mice where B cells respond to TLR4 stimulation, mature TLR4 protein in humans is only found on the surface of macrophages and dendritic cells and human B cells do not, therefore, respond to TLR4 agonists. LPS—when bound to the adaptor/transport proteins CD14 and MD-2—this is the most commonly described natural ligand of TLR4 (Takeuchi et al. 2001), but this receptor can also bind to a variety of PRR including: *Streptococcus pneumoniae* pneumolysin, *Chlamydia pneumoniae* HSP60, mouse mammary tumor virus-encoded envelope proteins, plant ligands, as well as several heat shock proteins (Takeuchi et al. 2001), components of the extracellular matrix such as hyaluronan, and fragments of fibronectin (O'Neill and Bowie 2007; Gribar et al. 2008; O'Neill et al. 2009; Bulut et al. 2005). TLR4-targeting adjuvants are the most advanced in clinical development with a few approved commercial vaccines developed by GlaxoSmithKline (GSK) for Hepatitis B ("Fendrix®") and for human papilloma virus, "Cervarix®". GSKs TLR4 ligand, MPL® or monophosphoryl lipid A, was developed by Ribi and Corixa Corp, and is a detoxified bacterial lipopolysaccharide which is produced by isolation and refinement from *Salmonella minnesota* (Persing et al. 2002; Vandepapeliere et al. 2008). When properly formulated, this agonist can elicit humoral and cellular responses characterized by strong antibody production and a Th1-bias. The commercial success and pipeline with MPL support the hypothesis that TLR4 agonists can be used for improved modern vaccines. Second-generation molecules that are synthetic and target TLR4 are also being developed. These are present in some cases as pure hexa-acylated molecules that can efficiently bind to human TLR4 (Rallabhandi et al. 2008). Due to their synthetic nature, they should be straightforward to scale for global production and eliminate some of the batch-to-batch variability inherent to biological products, as well as reduce the risk of potential off-target activities or undesirable effects of contaminants that are difficult to control in purified biological products. Some of these—like the synthetic agonists GLA and SLA—are being developed by IDRI and have shown promise in preclinical and human clinical studies (Anderson et al. 2010; Coler et al. 2011; Lousada-Dietrich et al. 2011). GLA is a synthetic analog of Lipid A and made based simply on the structure of the natural molecule (Lousada-Dietrich et al. 2011; Coler et al. 2011). After the structure of the human MD2/TLR4 complex was published (Park et al. 2009), rational inspection of the structure enabled predictions regarding the chain lengths and binding interactions required for a tight fit into the lipid A-binding pocket. The resulting molecule, SLA, has several desirable properties and has now been developed through human clinical trials (Baldwin et al. 2016; Paes et al. 2016). In these trials, the SLA molecule was able to elicit more robust adaptive responses in humans than similar TLR4 agonists, and it has become a lead candidate for some human vaccine indications (Fox et al. 2016; Carter et al. 2016).

TLR5

TLR5 is expressed by neutrophils, monocytes, T cells, and endothelial cells (Gribar et al. 2008). This TLR is the only known one that binds primarily to a protein target, the D1 region of bacterial flagellin (Hayashi et al. 2001). The ability to make fusions to the protein enables the expression vectors that couple antigen to the protein and resulting constructs auto-adjuvant by inducing inflammatory responses that include interferon production and enhanced adaptive humoral responses (Applequist et al. 2005). Co-delivery of antigen and TLR5 agonist in such a chimeric fusion protein yields immune responses superior to the analogous mixture of antigen and agonist. However, TLR5 engagement has been reported to increase antibody production against flu when intranasal influenza vaccines were given in animal models (Gupta et al. 2004; Otero et al. 2004). Widespread use of flagellin as a vaccine adjuvant maybe restricted, however, by preexisting or vaccine generated anti-flagellin immune responses. Nonetheless, commercial efforts are underway to use this agonist in commercial vaccines (Song et al. 2014).

TLR7 and TLR8

TLR7 and TLR8 are localized on endosomes within numerous cell types (O'Neill et al. 2009; Dasari et al. 2005) where they recognize RNA segments enriched for GU or poly-U sequences. While the natural ligand for these TLRs appears to be nucleic acid, a class of compounds, the imidazoquinolines, was also discovered to bind TLR7/8. The imidazoquinolines are straightforward to synthesize and therefore are attractive adjuvant components as they are inexpensive, stable, and amenable to multiple formulations. Members of these include the small molecules imiquimod and resiquimod which initiate strong immune responses (Smith et al. 2003;

Shukla et al. 2010), and which have been used commercially in host-directed therapy for years in the form of Aldara® cream (Gupta et al. 2004; Chollet et al. 1999). Extrapolating from murine studies to humans for some of these compounds is difficult as mice apparently lacks a functional TLR8 (Hemmi et al. 2002; Jurk et al. 2002; Gorden et al. 2006; Liu et al. 2010). Nonetheless, studies in mice suggest that subcutaneous or intramuscular immunization with either of these TLR7/8 agonists enhances the immune response to co-administered antigens (Otero et al. 2004; Zhang and Mosser 2008). Their solubility may hinder localization to the injection site, however, thereby reducing availability in the draining lymph node and the innate help is required to efficiently recognize the co-administered antigen. Topical pre-administration of these agonists could circumvent this problem, and recent human trials demonstrated that pre-administration of imiquimod can dramatically enhance immune responses to flu vaccines administered hours after dosing with the adjuvant (Hung et al. 2016; Hung et al. 2014). The other strategy to better utilize the imidazoquinolines in adjuvant formulations involves directly attaching a lipid tail, thereby slowing their passage and enabling local innate immune stimulation. This has recently been achieved by 3M and opens the window to new 3M052-based adjuvant formulations targeting TLR7 and TLR8 (Smirnov et al. 2011; Singh et al. 2014; Zhao et al. 2014; Van Hoeven et al. 2017).

TLR9

As with the other nucleic acid sensors TLR7 and TLR8, TLR9 is also located in the endosome (Dasari et al. 2005) where it recognizes desoxyribonucleotides, specifically, unmethylated CpG OligoDesoxyriboNucleic acids (CpG ODN) (Hemmi et al. 2003). CpG ODN are common within viral and bacterial genomes but is not generally found in mammalian genomes, making them good innate patterns for nonself discrimination (O'Neill et al. 2009). When properly formulated with antigens, this class of compounds can also enhance immune responses in animal models and humans. There are several classes of CpG that all stimulate TLR9-dependent signaling, although each with varied dosing and response relationships. The first, "A-class" ODN promote IFN-α secretion by plasmacytoid DC and the highest degree of NK cell stimulation; the second, "B-class" ODN stimulates strong cytokine production, B cell, and NK cell activation; and finally, the third, "C-class" ODN combine the effects of A- and B-class CpG ODN. Therefore, C-class ODN can be made into potent Th1-biasing adjuvants and can be used as therapeutic drugs for certain infectious disease or cancers (Zhao et al. 2014; Ali et al. 2014). Administration of CpG ODN via the pulmonary route has proven to be safe in primates and humans for treatment of allergic indications (Campbell et al. 2009). CpG ODN is being developed for commercial use by companies like Coley Pharmaceuticals (now Pfizer) and Dynavax Technologies, amongst others. Human trials with one of these compounds, a B-class ODN, CpG7909, demonstrated that these compounds elicit the desired innate immune responses with a pattern of Th1 cytokine and chemokine secretion resulting in altered trafficking of neutrophils, monocytes, and other lymphocytes in the blood, patterns which were reproduced upon a second injection with the agonist (Krieg et al. 2004). Another trial with an

ODN, CpG 10101 or "Actilon" demonstrated the safety of the agonist with doses as high as 20 mg given twice as well as with longer term administration of 4 mg given twice weekly for a total of eight injections (Halperin et al. 2003). Both trials confirmed the safety of administration of the TLR9 agonists although—as to be expected by their modes of action—there were local reactions at the injection sites and the elicited cytokines produced flu-like symptoms in the trial subjects. After a long period of regulatory scrutiny, Dynavax's Heplisav-B vaccine for hepatitis B consisting of recombinant hepatitis B surface antigen adjuvanted by CpG 1018 ISS was approved in the US in 2017 making TLR9 agonists one of the few US-approved adjuvant components (Campbell 2017; Lowes 2017).

Combinations of TLR agonists

Since TLR can act through two independent signaling pathways, MyD88 and TRIF, this raises the possibility that using multiple agonists that trigger two or more receptors might enhance immune responses by concurrently engaging both pathways (Kasturi et al. 2011; Carter and Reed 2010). TLR4 is known to recruit both MyD88 and TRIF, and—for example—monocyte-derived DC can integrate both TRIF and MyD88 signals received via different TLR. Indeed, triggering TLR3 with poly I:C or engaging TLR4 while stimulating TLR7 and TLR8 produces synergistic cytokine secretion as seen in enhanced IL-12p70 production (Bohnenkamp et al. 2007). This results in superior CD8+ T-cell responses induced by synergistically activated DC and is superior to engagement of each receptor alone, demonstrating improved function. More extensive screening on human PBMC with combinations of TLR ligands and looking at the resulting cytokine profiles has been performed (Ghosh et al. 2006). In these studies—consistent with the above—the TLR7 and TLR8 combinations with TLR4 agonists induced a strong Th1-biased response with enhanced IFN-γ, IL-12, and IFN-α but a more muted induction of TNF. Similarly, TLR7 and TLR8 combined with TLR2-targeted agonists synergistically up-regulated IFN-γ but interestingly, IL-12 secretion was not enhanced. In contrast, while TLR9 targeting generated robust interferon responses alone, the addition of TLR7/TLR8 agonists reduced TLR9-induced type 1 IFN. Finally, TLR3 targeting induced significant IFN-α/-β which was additive to that induced with agonists of TLR2, TLR5, TLR7, and TLR8 (Ghosh et al. 2006). This combination of adjuvants can enhance protective responses in vivo as we reported (Raman et al. 2010). Here, it was revealed that although a variety of TLR ligand combinations enhanced the numbers of specific IFN-γ-producing CD4 positive T cells, only the adaptive responses that followed combination TLR triggering using TLR4 and TLR9 agonists in a vaccine formulation with a *Leishmania* antigen candidate provided highly protective responses in mice to challenge with *L. major*. A better understanding of the subtle differences in cytokine compositions induced by synergistic TLR agonists may allow for tailored mixtures capable of producing very directed responses, decreasing off-target effects, and enhancing safe and powerful immunity.

3 Formulations

The innate immune system is responsive to ligands that are presented in a certain context, and this explains the value of formulating adjuvants in a manner consistent with the biological presentation. The synergy of inflammatory and TLR ligands, and the importance of reaching subcellular destinations of the ligand-mediated by particle composition and size are all key to the magnitude and quality of the immune response generated (Desbien et al. 2015; Misquith et al. 2014). Therefore, while instructive, simply mixing innate factors with antigens is a suboptimal approach to adjuvant development. Instead, careful attention to the desired responses, the biology of antigen/ adjuvant uptake and processing, as well as bringing the appropriate cell types into contact with each other, will result in more durable and effective vaccination outcomes (Orr et al. 2013; Fox et al. 2013). Here, is when appropriately formulating the agonist mixture truly renders an adjuvant and—when partnered with an appropriate antigen—forms the basis for an effective vaccine.

Vaccine antigens and adjuvants can be made more effective by employing appropriate particulate physical states to increase cell uptake and provide sustained release of antigen and the active pharmaceutical ingredients such as TLR agonists. Particulate formulations such as liposomes, aluminum gels, and aqueous suspensions are more amenable to uptake by APCs which specialize in the phagocytosis of invading particulate pathogens. When formulated properly, as presented below, these adjuvants appear capable of eliciting and boosting potent human immune responses in a safe manner.

Liposomes, emulsions, aluminum salts, and suspensions are formulations that maintain particles which are amenable to uptake by APC. An example of this are formulations of TLR4 agonists that contain phospholipids to form liposomes, emulsions, and suspensions of Lipid A-like molecules, which then as aggregates are more effective innate stimulants than soluble monomeric forms would be (Mueller et al. 2004). The specific geometry and phase in which the formulated molecule is presented dictate its biological activity (Mueller et al. 2004; Seydel et al. 2000). Certain lipid combinations—when added to the formulation—trigger IL-18 secretion and guide the responding cellular subsets resulting in appropriate immunity to the vaccine formulation (Desbien et al. 2015). In the following paragraphs, we describe some of the presentations used for adjuvant formulation:

3.1 Aqueous Formulations

Aqueous formulations are generally some of the simpler types to use. The agonist is usually placed in a buffered, aqueous base which is then used as the carrier for the innate agonist. Amphiphilic molecules, like many of the TLR4 and TLR2 agonists, will self-assemble into aggregates upon exposure to these aqueous solutions

(Fox et al. 2013). Manufacturing processes that include high energy input with sonication or microfluidization in the presence of small amounts of surfactant enable particle size reduction until a nanosuspension of aggregates is achieved. In some cases, these aqueous formulations are effective at inducing robust innate immune responses with adaptive immune responses equivalent to those elicited by more complex formulations such as those described below.

3.2 Emulsions

Emulsions have been shown to effectively and safely induce immune responses to influenza antigens and importantly, can enable dose sparing of the antigens (Reed Steven et al. 2012; Nitayaphan et al. 2000; Li et al. 2006). Oil-in-water (o/w) emulsions consist mostly of metabolizable oils emulsified with biocompatible surfactants in an aqueous bulk phase. Emulsion droplets are ~ 100 nm in diameter and are stable for years (Reed Steven et al. 2012). At scale, these emulsions are produced by microfluidization of the oil into the aqueous phase with a broad choice of oils that can be formulated without changing the desired size and stability properties dramatically when appropriately paired with an emulsifier (Fox 2009; Fox et al. 2010). The addition of these stable emulsions to vaccine formulations generally stimulates antibody responses but, without the addition of further stimulants, usually does not improve T-cell immunity (Guy 2007; Kwissa et al. 2007; Mosca et al. 2008; Fox et al. 2008). To enhance cellular responses, TLR agonists can be added to the emulsion producing potent adjuvants like the TLR4 agonist-containing squalene emulsion GLA-SE (Santini-Oliveira et al. 2016; Coler et al. 2015). In this presentation, amphiphilic molecules like TLR agonists localize at the oil/water interface (Anderson et al. 2010). The squalene emulsions like Novartis' MF59 and GSKs AS03 have been used in numerous presentations, and have been approved as commercial products for use in influenza vaccines. In the adjuvanted seasonal flu vaccine, Fluad®, it is thought that the MF59 oil-in-water emulsion works by causing release of extracellular ATP (Vono et al. 2013), prolonging APC survival, and making antigen processing more efficient (Seubert et al. 2008; Hamilton et al. 2000) (Fig. 1).

3.3 Alum-Based Formulations

Aluminum hydroxide or phosphate salts, often collectively referred to as "Alum", are one of the most commonly used adjuvants in commercial products and have been a mainstay of vaccine development for many years (Wen and Shi 2016). The alum particles consist of nanometer-sized crystals that assemble into aggregates of several μms and provide a stable particulate formulation for the sustained release of molecules adsorbed to the surface. The salts themselves have been studied by many

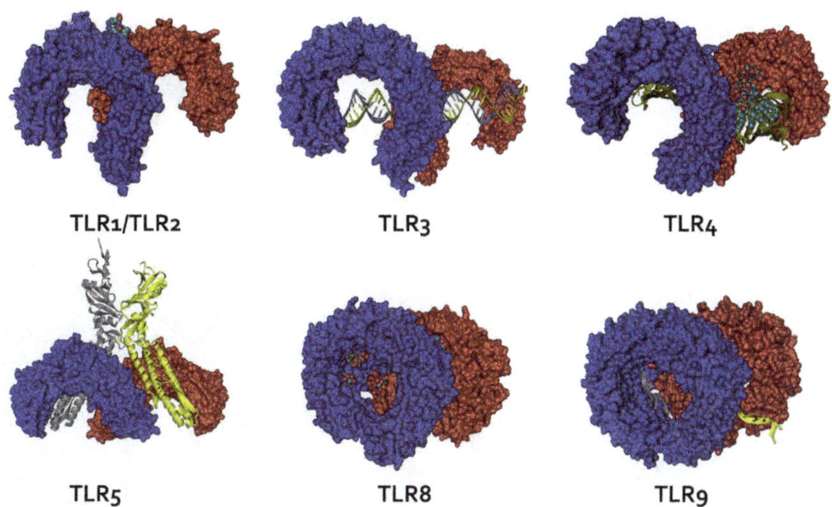

Fig. 1 Sample Structures of TLR. Shown are representative X-ray crystal structures of TLR in their active, ligand-bound form from various organisms. The remarkable structural conservation combined with the diversity of ligand is apparent in these structures. The TLR dimers are shown with each individual monomer shown blue or red. CPK color representation for small ligands and yellow/silver representation for large ones. TLR1/2—glycopeptide: The human molecules ectodomain is shown bound to Pam3CSK4 (Jin et al. 2007). TLR3—dsRNA: The murine TLR3 dimer ectodomain is depicted bound to dsRNA (Liu et al. 2008). TLR4—LPS: The human receptor in complex with MD2 is shown with LPS bound (Park et al. 2009). TLR5—flagellin: Zebrafish TLR5 is shown bound to flagellin from Salmonella (Yoon et al. 2012). TLR8—ssRNA: Human TLR8 was crystalized in the presence of uridine and an oligo ssRNA (Tanji et al. 2015). TLR9—CpG: The horse TLR9 dimer bound to oligo CpG DNA is shown here (Ohto et al. 2015)

groups, but the mechanism of action has remained elusive and incompletely defined. Reports have indicated that inflammatory mechanisms involving NLRP3 are part of the activity, but the "depot effect" and enhanced antigen uptake due to the affinity of APC membranes to the salts have all been invoked to explain Alum's activity. No single explanation of alum's activity has emerged, and it may well be that multiple mechanisms are at work (Spreafico et al. 2010; He et al. 2015; Ruwona et al. 2016). To expand on its ability to initiate immune responses and target certain cell populations, one can adsorb agonists of the innate immune system onto aluminum hydroxide (alum) particles (Chuai et al. 2013) and certain TLR4 agonists, like MPL and GLA, strongly adsorb to alum and direct the quality of the immune response to a balanced Th1/Th2 response, whereas alum alone induces a Th2 response (Beran 2008). This has led to the development of the adjuvant AS04 by GSK (Didierlaurent et al. 2009), and in the form of the Cervarix[®] papilloma virus vaccine, an alum-adsorbed MPL which is the first defined TLR4-based adjuvant to be approved in a commercial vaccine in the United States (Schwarz 2009) (Fig. 2).

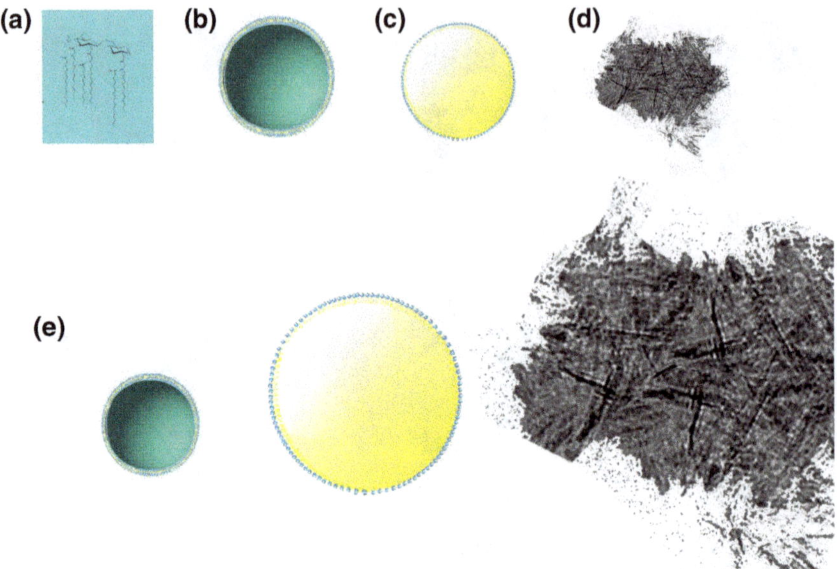

Fig. 2 Formulation Configuration. **a** *A hydrophobic TLR agonist in water*; Aqueous presentation of the molecule is the simplest way to produce an adjuvant, however, this presentation is usually not optimal for robust immune responses (Fox et al. 2013). **b** *A liposome with a lipid bilayer enclosing the interior*; The liposome offers many advantages including the ability to co-present multiple ligands as well as incorporating some in the interior of the liposome, the surface of it, and in the space between liposomes (Fox et al. 2014). **c** *An oil-in-water emulsion droplet*; Here, the oil itself can become part of the immune stimulating components. Agonists can be incorporated into the droplet allowing potent immune stimulation (Behzad et al. 2012; Lofano et al. 2015). **d** *Alum crystals*; alum salts can adsorb TLR agonists and are themselves one of the most well-known adjuvants having been used for over 100 years (Wen and Shi 2016; Didierlaurent et al. 2009). **e** *Size comparison*. Here, the formulation choices from the top row have been scaled to their relative sizes. The ligands themselves are usually very small on the scale of 1–5 nm while alum salts can have micrometer-sized crystals in them

3.4 Liposomes

Liposomes consist of vesicles often formed by the assembly of phospholipid bilayers and are versatile carriers of innate signaling molecules (Vandepapeliere et al. 2008; Gram et al. 2009; Steers et al. 2009). For optimal uptake, these vesicles are produced to be around 100 nm in diameter, and can be optimized to be stable for years under proper storage conditions. Liposomes are versatile in that they can carry payloads in their interior as well as on their surface and, due to the choice of excipients used to produce them, are biocompatible vehicles for adjuvant delivery that can dramatically enhancing immune responses (Baldwin et al. 2016; Orr et al. 2013). Amphiphilic molecules like certain TLR agonists are likely localized within the lipid bilayer of the vesicle while aqueous components that can add stimulatory capacity can be encapsulated in the aqueous interior of the liposomes or solubilized in the exterior bulk phase (Fox et al. 2014). One of the most advanced liposomal

Table 1 TLR and their ligands

Receptor	Architecture	Major ligand class	Sample ligand	Putative target sensing
Plasma membrane				
TLR1	TLR1/TLR2	Tri-acylated Lipopeptides	Pam3CSK4	Bacterial surface components
TLR2	TLR1,2, or 6 dimer	Lipopeptides	Pam2CSK4	
TLR6	TLR2/TLR6	Diacylated Lipopeptides	FSL-1	
TLR4	(TLR4/MD2) dimer	Lipopolysaccharides	MPL	Bacterial membrane glycolipids
TLR5	TLR5 dimer	Flagellins	*E. coli* flagellin	Flagella from pathogens
TLR9	TLR9 dimer	Unmethylated DNA	ODN CpG	Bacterial DNA
Endosomal				
TLR3	TLR3 dimer	Double-stranded RNA	Poly(I:C)	Viral nucleic acids
TLR7	TLR7 dimer	Single-stranded RNA	Imiquimod	Viral nucleic acids
TLR8	TLR8 dimer	Single-stranded RNA	Guardiquimod	Viral nucleic acids

adjuvant, AS01, is a formulation of the TLR4 agonist MPL with the saponin QS21. AS01 has been used in licensure trials of a new malaria vaccine, RTS,S (Ansong et al. 2011; Leroux-Roels et al. 2013). By combining the signals induced by the saponin QS21 with the innate signaling of TLR4, a potent immune response is elicited in numerous species—including nonhuman primates—and this response appears to be much more durable than that induced by many other adjuvant/antigen combinations (Hay et al. 2014) (Table 1).

4 Enhancing Vaccines Through Adjuvant Action

Using tools to combine the agonists targeting various receptors of the innate immune system as described above into formulations that provide optimal presentation and compatibility with the target antigen, one can develop adjuvant formulations that enhance various aspects of the immune response and generate powerful adaptive immunity (Table 2).

4.1 Protecting Vulnerable Populations

Adjuvant development considerations for vaccines that should protect the adult population may differ from those required for targeting of vulnerable subsets like the very young or the elderly. Many standard vaccines are routinely given in the

Table 2 Enhancing vaccines using adjuvants

Effect	Explanation	Examples of solutions with adjuvants
Protecting vulnerable populations	Some populations such as HIV/AIDS patients, the elderly, or the very young do not respond strongly to standard vaccine formulations	By adding an adjuvant before injection, Hung et al. were able to drastically increase protection in the elderly (Hung et al. 2016; Hung et al. 2014)
Broadening the immune response	Vaccines without appropriate innate help may only protect against a single strain of pathogen causing failure for some vaccines or necessitating constant reformulation like the flu vaccine	Stimulating innate responses increased the breadth of the immune response against a malaria antigen allowing for neutralizing responses against other serovars of the parasite (Wiley et al. 2011; Matthews et al. 2013)
Providing long-term protection	While short-term immune responses could ward off infections, without appropriate innate stimulation long-term immunity may not be achieved	Martins et al. studied the durability of protection afforded by a VLP to a mouse-adapted Ebola challenge. While protection was robust early on, after 22 weeks, the protection waned. This was reversible upon inclusion of innate stimulants and appeared to correlate with the induction of CD4 responses (Martins et al. 2016)
Dose sparing	For some vaccines, antigen supply can be limited; therefore, strategies to stretch the supply can help with costs and population coverage, important considerations for example, pandemic flu vaccines	In studies in humans using a pandemic H7 formulation, Mulligan et al. measured negligible titers (5% seroconversion) in the high dose group receiving 45 µg of antigen, while at the lowest dose tested with adjuvant of 3.75 µg, there seroconversion rates were over ten times higher (Mulligan et al. 2014)
Dosage sparing	Rapid onset of responses and the ability to protect people with fewer immunizations increases effectiveness and compliance to vaccine regimens. Additionally, by decreasing the number of times a patient needs to be immunized, the cost to health care rapidly decreases	When given with hepatitis B vaccines, adjuvants potently increase seroconversion rates even after fewer doses. This was demonstrated both by GSK with the two-shot Fendrix vaccine and by Dynavax with inclusion of their CpG-based adjuvant (Beran 2008; Surquin et al. 2011; Halperin et al. 2012)

first year of life, a time when responses to TLR ligands differs from those seen in older individuals. For example, blood from adults is more likely to respond with Th1-type cytokines like IL-12p70 and IFN-α than was cord blood. As the newborn matures, responses rapidly climb to be more similar to adults and responses to TLR3, TLR7, and TLR9 ligands are more adult-like than those found to TLR4 ligands where high levels of IL-10 are elicited early in life (Belderbos et al. 2009). By 6–9 months of age, TLR4 responses begin to reach adult levels while even one year into life TLR9 agonist responses lag when compared to adults (Nguyen et al. 2010). Therefore, when developing adjuvants for vaccines targeted to neonates, one needs to consider the combinations of agonists capable of eliciting the desired immune responses.

The responses to innate ligands late in life are the mirror image to those observed in the early stages: while early in life responses rapidly rise, later in life, there is a gradual waning of responsiveness to innate agonists. Due to the advanced nature of TLR4 agonists these have been the most studied, but inconsistent observations have been made in the elderly: data seem to indicate that later in life TLR4 agonism is largely preserved—especially, when presented in the context of a squalene emulsion (van Duin and Shaw 2007; Behzad et al. 2012). Clearer data are available for agonists of other TLR: for example, cytokine production from monocytes in the elderly was substantially decreased in response to TLR2 agonists and upregulation of costimulatory molecules on monocytes and lower cytokine production from DC was reported for a variety of TLR ligands (van Duin et al. 2007; Panda et al. 2010). These impaired responses likely contribute to susceptibility of the elderly to infection and poor protective responses in some cases to flu and Lyme disease vaccines in adults over 60 (Alexopoulou et al. 2002; Sigal et al. 1998). Finally, other immune-compromised patients could benefit from the addition of innate stimulants into vaccine preparations. Results from trials with hepatitis B and pneumococcal conjugate vaccines have demonstrated that inclusion of the TLR9 agonist CpG7909 was able to boost responses and increase the percentage of durable responders in HIV-infected individuals and the elderly (Sogaard et al. 2010). For these reasons, the selection of appropriate innate stimulants formulated as appropriate adjuvants requires careful consideration of the underlying biology of the target population. When achieved, this can overcome hypo-responsiveness in vulnerable subsets of the population.

4.2 Broadening the Immune Response

Many pathogens present with variations that make it difficult for a focused immune response to neutralize antigens that are "drifted" from those found in vaccine preparations. This can allow strains of a pathogen to escape the immunity induced to any given serovar. A familiar example of this is the annual emergence of newly circulating flu strains and the resulting need to predict the strains that will infect in any given season and adjust the vaccine formulation accordingly. To address this, vaccines that can induce responses both to the antigen in the vaccine, but also have these responses "broadened" so that they can recognize slight variations that may be present in drifted pathogen strains, would be ideal. Immunization with formulations containing TLR agonist and antigens causes induction of interferon response genes, ligation of the B cell receptor, and activation of NFκB elements leading to class switching, somatic hypermutation, and concomitant affinity maturation against the target (Pone et al. 2012). While class switching is important for the production of antibody isotypes that, e.g., efficiently fix complement, the action of Activation-Induced Cytidine Deaminase (AID) additionally introduces mutations in variable regions of the antibody by changing deoxycytidine to deoxyuridine, which is then dealt with through DNA repair enzymes resulting in point mutations that can change amino acid sequences in the binding regions of the antibody (Li et al. 2004; Brenner and Milstein 1966). B cells of higher affinity are then selected out during germinal center reactions in a process of survival of the fittest (Gitlin et al. 2014). While higher avidity antibodies emerge from this process, it is now also known that the diversity of antibody sequences that recognizes an antigen also increases (Kasturi et al. 2011). Instead of simply being a result of enhanced affinity of a fixed number of clones, the diversity of new variable sequences greatly increases when TLR ligands are present allowing recognition of drifted pathogens as reported by Wiley et al. for the malaria vaccine PvRII (Wiley et al. 2011). The inclusion of TLR agonists into adjuvanted vaccines for vivax malaria greatly increased the diversity of variable chain sequences in immunized animals resulting in recognition of proteins from malaria strains not included in the vaccine preparation. This is an important finding with implications for both established and emerging diseases where immunity to single serovars will not be sufficient for population-wide protection.

4.3 Providing Long-Term Protection

While the initial induction of protection is the most important property of a vaccine, with the exclusion of explosive epidemics, prophylactic vaccines that do not provide at least some long-term protection are doomed to fail since more than annual re-immunization for a single disease is not sustainable. For this reason, subjects need to respond to the vaccine with a protective response that persists rather than immediately waning. The inclusion of adjuvants may be a way to tune the immune response and lead to such lasting immunity (Bergmann-Leitner and Leitner 2014). An interesting example of this may be the first successful vaccine trial Rv144 or the "Thai trial". Here, the groups that were immunized differed significantly in the first months after vaccination, but this "blip" of protection then waned and both groups followed parallel curves for the rest of the trial (Excler et al. 2013). It is tempting to hypothesize that an appropriate quality response was generated early on but that it was just not durable enough to provide sustained protection. Since CD4 T cells may be critical for long-lived responses (Martins et al. 2016) and are the target for HIV infection, a balance of the two could be required to provide an effective long-lived response for HIV (Lewis et al. 2014). Immunizing with emulsion or alum-based adjuvants alone may not be sufficient for long-lived immunity and either prime-boost vaccination or the inclusion of TLR agonists in formulations appears to be required for a long-lived immune response (Martins et al. 2016; Leroux-Roels et al. 2015; Lofano et al. 2015).

4.4 Dose and Dosage Sparing

Adding adjuvants to vaccine preparations can allow both dose sparing, i.e., the use of less of an antigen for which limited supplies may be available; and dosage sparing, the use of fewer vaccine injections or dosings required to achieve protective immunity. Related to this is the ability to accelerate responsiveness. In a demonstration of this in human trials, the TLR9 agonist CpG7909 was added to a commercial hepatitis B vaccine. Antigen-specific responses were seen much earlier, and these responses were considerably higher than in subjects receiving the vaccine without the added innate agonist. Most subjects already had measurable titers 2 weeks after a prime and some exhibited cytotoxic cellular responses in the higher CpG dose groups (Halperin et al. 2006). After a single boost, most subjects were seropositive for HBsAg antibodies demonstrating the feasibility of accelerated seroconversion (Sogaard et al. 2010; Tighe et al. 2000).

5 Conclusion

With the vaccine development field moving into a modern era of defined vaccines that are produced in a controlled and reproducible manner to provide safe and controlled protection to vast populations, there is increasing demand to provide adjuvant formulations that maintain the potency of these vaccines while allowing for targeted immune stimulation. Modern, defined adjuvants are entering the toolbox of vaccine developers, and will likely be critical in protecting vulnerable populations and the rest of the world from diseases for which vaccines have previously been thought beyond reach. An emerging concept is that these will likely be combination adjuvants with the potential to trigger multiple innate pathways to provide targeted and high-quality immune responses.

References

Alexopoulou L, Holt AC, Medzhitov R, Flavell RA (2001) Recognition of double-stranded RNA and activation of NF-kappaB by Toll-like receptor 3. Nature 413:732–738
Alexopoulou L, Thomas V, Schnare M, Lobet Y, Anguita J, Schoen RT, Medzhitov R, Fikrig E, Flavell RA (2002) Hyporesponsiveness to vaccination with *Borrelia burgdorferi* OspA in humans and in TLR1- and TLR2-deficient mice. Nat Med 8:878–884

Ali OA, Verbeke C, Johnson C, Sands W, Lewin SA, White D, Doherty E, Dranoff G, Mooney DJ (2014) Identification of immune factors regulating anti-tumor immunity using polymeric vaccines with multiple adjuvants. Cancer Res

Anderson RC, Fox CB, Dutill TS, Shaverdian N, Evers TL, Poshusta GR, Chesko J, Coler RN, Friede M, Reed SG et al (2010) Physicochemical characterization and biological activity of synthetic TLR4 agonist formulations. Colloids Surf B: Biointerfaces 75:123–132

Ansong D, Asante KP, Vekemans J, Owusu SK, Owusu R, Brobby NA, Dosoo D, Osei-Akoto A, Osei-Kwakye K, Asafo-Adjei E et al (2011) T cell responses to the RTS, S/AS01(E) and RTS, S/AS02(D) malaria candidate vaccines administered according to different schedules to Ghanaian children. PLoS ONE 6:e18891

Applequist SE, Rollman E, Wareing MD, Liden M, Rozell B, Hinkula J, Ljunggren HG (2005) Activation of innate immunity, inflammation, and potentiation of DNA vaccination through mammalian expression of the TLR5 agonist flagellin. J Immunol 175:3882–3891

Baldridge JR, McGowan P, Evans JT, Cluff C, Mossman S, Johnson D, Persing D (2004) Taking a Toll on human disease: Toll-like receptor 4 agonists as vaccine adjuvants and monotherapeutic agents. Expert Opin Biol Ther 4:1129–1138

Baldwin SL, Roeffen W, Singh SK, Tiendrebeogo RW, Christiansen M, Beebe E, Carter D, Fox CB, Howard RF, Reed SG et al (2016) Synthetic TLR4 agonists enhance functional antibodies and CD4+ T-cell responses against the *Plasmodium falciparum* GMZ2.6C multi-stage vaccine antigen. Vaccine 34:2207–2215

Basith S, Manavalan B, Lee G, Kim SG, Choi S (2011) Toll-like receptor modulators: a patent review (2006–2010). Expert Opin Ther Pat 21:927–944

Behzad H, Huckriede ALW, Haynes L, Gentleman B, Coyle K, Wilschut JC, Kollmann TR, Reed SG, McElhaney JE (2012) GLA-SE, a synthetic toll-like receptor 4 agonist, enhances T-cell responses to influenza vaccine in older adults. J Infect Dis 205:466–473

Belderbos ME, van Bleek GM, Levy O, Blanken MO, Houben ML, Schuijff L, Kimpen JL, Bont L (2009) Skewed pattern of Toll-like receptor 4-mediated cytokine production in human neonatal blood: low LPS-induced IL-12p70 and high IL-10 persist throughout the first month of life. Clin Immunol 133:228–237

Beran J (2008) Safety and immunogenicity of a new hepatitis B vaccine for the protection of patients with renal insufficiency including pre-haemodialysis and haemodialysis patients. Expert Opin Biol Ther 8:235–247

Bergmann-Leitner ES, Leitner WW (2014) Adjuvants in the driver's seat: how magnitude, type, fine specificity and longevity of immune responses are driven by distinct classes of immune potentiators. Vaccines (Basel) 2:252–296

Bohnenkamp HR, Papazisis KT, Burchell JM, Taylor-Papadimitriou J (2007) Synergism of Toll-like receptor-induced interleukin-12p70 secretion by monocyte-derived dendritic cells is mediated through p38 MAPK and lowers the threshold of T-helper cell type 1 responses. Cell Immunol 247:72–84

Breckpot K, Escors D, Arce F, Lopes L, Karwacz K, Van Lint S, Keyaerts M, Collins M (2010) HIV-1 lentiviral vector immunogenicity is mediated by TLR3 and TLR7. J Virol

Brenner S, Milstein C (1966/2006) Pillars article: origin of antibody variation. Nature 211:242–243. J Immunol 177:4237–4238

Bretscher P, Wei G, Menon J, Bielefeldt-Ohmann H (1992) Establishment of stable, cell-mediated immunity that makes "susceptible" mice resistant to Leishmania major. Science 257:539–542

Bulut Y, Michelsen KS, Hayrapetian L, Naiki Y, Spallek R, Singh M, Arditi M (2005) Mycobacterium tuberculosis heat shock proteins use diverse Toll-like receptor pathways to activate pro-inflammatory signals. J Biol Chem 280:20961–20967

Campbell JD (2017) Development of the CpG adjuvant 1018: a case study. Methods Mol Biol 1494:15–27

Campbell JD, Cho Y, Foster ML, Kanzler H, Kachura MA, Lum JA, Ratcliffe MJ, Sathe A, Leishman AJ, Bahl A et al (2009) CpG-containing immunostimulatory DNA sequences elicit TNF-alpha-dependent toxicity in rodents but not in humans. J Clin Invest 119:2564–2576

Capolunghi F, Rosado MM, Sinibaldi M, Aranburu A, Carsetti R (2013) Why do we need IgM memory B cells? Immunol Lett 152:114–120

Carter D, Reed SG (2010) Role of adjuvants in modeling the immune response. Curr Opin HIV AIDS 5:409–413. https://doi.org/10.1097/COH.1090b1013e32833d32832cdb

Carter D, Fox CB, Day TA, Guderian JA, Liang H, Rolf T, Vergara J, Sagawa ZK, Ireton G, Orr MT et al (2016) A structure-function approach to optimizing TLR4 ligands for human vaccines. Clin Trans Immunol 5:e108

Cataldi A, Yevsa T, Vilte DA, Schulze K, Castro-Parodi M, Larzabal M, Ibarra C, Mercado EC, Guzman CA (2008) Efficient immune responses against Intimin and EspB of enterohaemorragic *Escherichia coli* after intranasal vaccination using the TLR2/6 agonist MALP-2 as adjuvant. Vaccine 26:5662–5667

Chang ZL (2010) Important aspects of Toll-like receptors, ligands and their signaling pathways. Inflamm Res

Chollet JL, Jozwiakowski MJ, Phares KR, Reiter MJ, Roddy PJ, Schultz HJ, Ta QV, Tomai MA (1999) Development of a topically active imiquimod formulation. Pharm Dev Technol 4:35–43

Chuai X, Chen H, Wang W, Deng Y, Wen B, Ruan L, Tan W (2013) Poly(I:C)/alum mixed adjuvant priming enhances HBV subunit vaccine-induced immunity in mice when combined with recombinant adenoviral-based HBV vaccine boosting. PLoS ONE 8:e54126

Coler RN, Bertholet S, Moutaftsi M, Guderian JA, Windish HP, Baldwin SL, Laughlin EM, Duthie MS, Fox CB, Carter D et al (2011) Development and characterization of synthetic glucopyranosyl lipid adjuvant system as a vaccine adjuvant. PLoS ONE 6:e16333

Coler RN, Duthie MS, Hofmeyer KA, Guderian J, Jayashankar L, Vergara J, Rolf T, Misquith A, Laurance JD, Raman VS et al (2015) From mouse to man: safety, immunogenicity and efficacy of a candidate leishmaniasis vaccine LEISH-F3+GLA-SE. Clin Trans Immunol 4:e35

Dasari P, Nicholson IC, Hodge G, Dandie GW, Zola H (2005) Expression of toll-like receptors on B lymphocytes. Cell Immunol 236:140–145

Degn SE, Thiel S (2013) Humoral pattern recognition and the complement system. Scand J Immunol 78:181–193

Desbien AL, Reed SJ, Bailor HR, Cauwelaert ND, Laurance JD, Orr MT, Fox CB, Carter D, Reed SG, Duthie MS (2015) Squalene emulsion potentiates the adjuvant activity of the TLR4 agonist, GLA, via inflammatory caspases, IL-18, and IFN-γ. Eur J Immunol 45:407–417

Didierlaurent AM, Morel S, Lockman L, Giannini SL, Bisteau M, Carlsen H, Kielland A, Vosters O, Vanderheyde N, Schiavetti F et al (2009) AS04, an aluminum salt- and TLR4 agonist-based adjuvant system, induces a transient localized innate immune response leading to enhanced adaptive immunity. J Immunol 183:6186–6197

El Shikh ME, El Sayed RM, Wu Y, Szakal AK, Tew JG (2007) TLR4 on follicular dendritic cells: an activation pathway that promotes accessory activity. J Immunol 179:4444–4450

Excler JL, Tomaras GD, Russell ND (2013) Novel directions in HIV-1 vaccines revealed from clinical trials. Curr Opin HIV AIDS 8:421–431

Foldes G, Liu A, Badiger R, Paul-Clark M, Moreno L, Lendvai Z, Wright JS, Ali NN, Harding SE, Mitchell JA (2010) Innate immunity in human embryonic stem cells: comparison with adult human endothelial cells. PLoS ONE 5:e10501

Fox CB (2009) Squalene emulsions for parenteral vaccine and drug delivery. Molecules 14:3286–3312

Fox CB, Anderson RC, Dutill TS, Goto Y, Reed SG, Vedvick TS (2008) Monitoring the effects of component structure and source on formulation stability and adjuvant activity of oil-in-water emulsions. Colloids Surf B Biointerfaces 65:98–105

Fox CB, Lin S, Sivananthan SJ, Dutill TS, Forseth KT, Reed SG, Vedvick TS (2010) Effects of emulsifier concentration, composition, and order of addition in squalene-phosphatidylcholine oil-in-water emulsions. Pharm Dev Technol

Fox CB, Moutaftsi M, Vergara J, Desbien AL, Nana GI, Vedvick TS, Coler RN, Reed SG (2013) TLR4 ligand formulation causes distinct effects on antigen-specific cell-mediated and humoral immune responses. Vaccine 31:5848–5855

Fox CB, Sivananthan SJ, Duthie MS, Vergara J, Guderian JA, Moon E, Coblentz D, Reed SG, Carter D (2014) A nanoliposome delivery system to synergistically trigger TLR4 AND TLR7. J Nanobiotechnol 12:17

Fox C, Carter D, Kramer R, Beckmann A, Reed S (2016) Current status of Toll-like receptor 4 ligand vaccine adjuvants. In: Schijns V, O'Hagan D (eds) Immunopotentiators in modern vaccines, 2nd edn. Elsevier Academic Press, pp 105–128

Gallorini S, Berti F, Mancuso G, Cozzi R, Tortoli M, Volpini G, Telford JL, Beninati C, Maione D, Wack A (2009) Toll-like receptor 2 dependent immunogenicity of glycoconjugate vaccines containing chemically derived zwitterionic polysaccharides. Proc Natl Acad Sci U S A 106:17481–17486

Gawchik SM, Saccar CL (2009) Pollinex Quattro Tree: allergy vaccine. Expert Opin Biol Ther 9:377–382

Ghosh TK, Mickelson DJ, Fink J, Solberg JC, Inglefield JR, Hook D, Gupta SK, Gibson S, Alkan SS (2006) Toll-like receptor (TLR) 2-9 agonists-induced cytokines and chemokines: I. Comparison with T cell receptor-induced responses. Cell Immunol 243:48–57

Gitlin AD, Shulman Z, Nussenzweig MC (2014) Clonal selection in the germinal centre by regulated proliferation and hypermutation. Nature 509:637–640

Gorden KK, Qiu XX, Binsfeld CC, Vasilakos JP, Alkan SS (2006) Cutting edge: activation of murine TLR8 by a combination of imidazoquinoline immune response modifiers and polyT oligodeoxynucleotides. J Immunol 177:6584–6587

Gram GJ, Karlsson I, Agger EM, Andersen P, Fomsgaard A (2009) A novel liposome-based adjuvant CAF01 for induction of CD8(+) cytotoxic T-lymphocytes (CTL) to HIV-1 minimal CTL peptides in HLA-A*0201 transgenic mice. PLoS ONE 4:e6950

Gribar SC, Richardson WM, Sodhi CP, Hackam DJ (2008) No longer an innocent bystander: epithelial toll-like receptor signaling in the development of mucosal inflammation. Mol Med 14:645–659

Gupta AK, Cherman AM, Tyring SK (2004) Viral and nonviral uses of imiquimod: a review. J Cutan Med Surg 8:338–352

Guy B (2007) The perfect mix: recent progress in adjuvant research. Nat Rev Microbiol 5:505–517

Halperin SA, Van Nest G, Smith B, Abtahi S, Whiley H, Eiden JJ (2003) A phase I study of the safety and immunogenicity of recombinant hepatitis B surface antigen co-administered with an immunostimulatory phosphorothioate oligonucleotide adjuvant. Vaccine 21:2461–2467

Halperin SA, Dobson S, McNeil S, Langley JM, Smith B, McCall-Sani R, Levitt D, Nest GV, Gennevois D, Eiden JJ (2006) Comparison of the safety and immunogenicity of hepatitis B virus surface antigen co-administered with an immunostimulatory phosphorothioate oligonucleotide and a licensed hepatitis B vaccine in healthy young adults. Vaccine 24:20–26

Halperin SA, Ward B, Cooper C, Predy G, Diaz-Mitoma F, Dionne M, Embree J, McGeer A, Zickler P, Moltz K-H et al (2012) Comparison of safety and immunogenicity of two doses of investigational hepatitis B virus surface antigen co-administered with an immunostimulatory phosphorothioate oligodeoxyribonucleotide and three doses of a licensed hepatitis B vaccine in healthy adults 18–55 years of age. Vaccine 30:2556–2563

Hamilton JA, Byrne R, Whitty G (2000) Particulate adjuvants can induce macrophage survival, DNA synthesis, and a synergistic proliferative response to GM-CSF and CSF-1. J Leukoc Biol 67:226–232

Hasan U, Chaffois C, Gaillard C, Saulnier V, Merck E, Tancredi S, Guiet C, Briere F, Vlach J, Lebecque S et al (2005) Human TLR10 is a functional receptor, expressed by B cells and plasmacytoid dendritic cells, which activates gene transcription through MyD88. J Immunol 174:2942–2950

Hay J, Carter D, Lieber A, Astier AL (2014) Recombinant Ad35 adenoviral proteins as potent modulators of human T cell activation. Immunology

Hayashi F, Smith KD, Ozinsky A, Hawn TR, Yi EC, Goodlett DR, Eng JK, Akira S, Underhill DM, Aderem A (2001) The innate immune response to bacterial flagellin is mediated by Toll-like receptor 5. Nature 410:1099–1103

He P, Zou Y, Hu Z (2015) Advances in aluminum hydroxide-based adjuvant research and its mechanism. Hum Vaccin Immunother 11:477–488

Hemmi H, Takeuchi O, Kawai T, Kaisho T, Sato S, Sanjo H, Matsumoto M, Hoshino K, Wagner H, Takeda K et al (2000) A Toll-like receptor recognizes bacterial DNA. Nature 408:740–745

Hemmi H, Kaisho T, Takeuchi O, Sato S, Sanjo H, Hoshino K, Horiuchi T, Tomizawa H, Takeda K, Akira S (2002) Small anti-viral compounds activate immune cells via the TLR7 MyD88-dependent signaling pathway. Nat Immunol 3:196–200

Hemmi H, Kaisho T, Takeda K, Akira S (2003) The roles of Toll-like receptor 9, MyD88, and DNA-dependent protein kinase catalytic subunit in the effects of two distinct CpG DNAs on dendritic cell subsets. J Immunol 170:3059–3064

Hung IFN, Zhang AJ, To KKW, Chan JFW, Li C, Zhu H-S, Li P, Li C, Chan T-C, Cheng VCC et al (2014) Immunogenicity of intradermal trivalent influenza vaccine with topical imiquimod: a double blind randomized controlled trial. Clin Infect Dis 59:1246–1255

Hung IF-N, Zhang AJ, To KK-W, Chan JF-W, Li P, Wong T-L, Zhang R, Chan T-C, Chan BC-Y, Wai HH et al (2016) Topical imiquimod before intradermal trivalent influenza vaccine for protection against heterologous non-vaccine and antigenically drifted viruses: a single-centre, double-blind, randomised, controlled phase 2b/3 trial. Lancet Infect Dis 16:209–218

Imler JL, Zheng L (2004) Biology of Toll receptors: lessons from insects and mammals. J Leukoc Biol 75:18–26

Jin MS, Kim SE, Heo JY, Lee ME, Kim HM, Paik S-G, Lee H, Lee J-O (2007) Crystal structure of the TLR1-TLR2 heterodimer induced by binding of a tri-acylated lipopeptide. Cell 130:1071–1082

Johnson TR, Rao S, Seder RA, Chen M, Graham BS (2009) TLR9 agonist, but not TLR7/8, functions as an adjuvant to diminish FI-RSV vaccine-enhanced disease, while either agonist used as therapy during primary RSV infection increases disease severity. Vaccine 27:3045–3052

Jurk M, Heil F, Vollmer J, Schetter C, Krieg AM, Wagner H, Lipford G, Bauer S (2002) Human TLR7 or TLR8 independently confer responsiveness to the antiviral compound R-848. Nat Immunol 3:499

Kang DC, Gopalkrishnan RV, Wu Q, Jankowsky E, Pyle AM, Fisher PB (2002) mda-5: An interferon-inducible putative RNA helicase with double-stranded RNA-dependent ATPase activity and melanoma growth-suppressive properties. Proc Natl Acad Sci U S A 99:637–642

Kasturi SP, Skountzou I, Albrecht RA, Koutsonanos D, Hua T, Nakaya HI, Ravindran R, Stewart S, Alam M, Kwissa M et al (2011) Programming the magnitude and persistence of antibody responses with innate immunity. Nature 470:543–547

Krieg AM, Efler SM, Wittpoth M, Al Adhami MJ, Davis HL (2004) Induction of systemic TH1-like innate immunity in normal volunteers following subcutaneous but not intravenous administration of CPG 7909, a synthetic B-class CpG oligodeoxynucleotide TLR9 agonist. J Immunother 27:460–471

Kwissa M, Kasturi SP, Pulendran B (2007) The science of adjuvants. Expert Rev Vaccines 6:673–684

Leroux-Roels I, Forgus S, De Boever F, Clement F, Demoitie MA, Mettens P, Moris P, Ledent E, Leroux-Roels G, Ofori-Anyinam O (2013) Improved CD4(+) T cell responses to *Mycobacterium tuberculosis* in PPD-negative adults by M72/AS01 as compared to the M72/AS02 and Mtb72F/AS02 tuberculosis candidate vaccine formulations: a randomized trial. Vaccine 13:2196–2206

Leroux-Roels G, Van Belle P, Vandepapeliere P, Horsmans Y, Janssens M, Carletti I, Garcon N, Wettendorff M, Van Mechelen M (2015) Vaccine Adjuvant Systems containing monophosphoryl lipid A and QS-21 induce strong humoral and cellular immune responses against hepatitis B surface antigen which persist for at least 4 years after vaccination. Vaccine 33:1084–1091

Lewis GK, DeVico AL, Gallo RC (2014) Antibody persistence and T-cell balance: two key factors confronting HIV vaccine development. Proc Natl Acad Sci U S A 111:15614–15621

Li Z, Woo CJ, Iglesias-Ussel MD, Ronai D, Scharff MD (2004) The generation of antibody diversity through somatic hypermutation and class switch recombination. Genes Dev 18:1–11

Li Y, Svehla K, Mathy NL, Voss G, Mascola JR, Wyatt R (2006) Characterization of antibody responses elicited by human immunodeficiency virus type 1 primary isolate trimeric and monomeric envelope glycoproteins in selected adjuvants. J Virol 80:1414–1426

Liu L, Botos I, Wang Y, Leonard JN, Shiloach J, Segal DM, Davies DR (2008) Structural basis of toll-like receptor 3 signaling with double-stranded RNA. Science 320:379–381

Liu J, Xu C, Hsu LC, Luo Y, Xiang R, Chuang TH (2010) A five-amino-acid motif in the undefined region of the TLR8 ectodomain is required for species-specific ligand recognition. Mol Immunol 47:1083–1090

Lofano G, Mancini F, Salvatore G, Cantisani R, Monaci E, Carrisi C, Tavarini S, Sammicheli C, Rossi Paccani S, Soldaini E et al (2015) Oil-in-water emulsion MF59 increases germinal center B cell differentiation and persistence in response to vaccination. J Immunol 195:1617–1627

Lousada-Dietrich S, Jogdand PS, Jepsen S, Pinto VV, Ditlev SB, Christiansen M, Larsen SO, Fox CB, Raman VS, Howard RF et al (2011) A synthetic TLR4 agonist formulated in an emulsion enhances humoral and Type 1 cellular immune responses against GMZ2—a GLURP-MSP3 fusion protein malaria vaccine candidate. Vaccine 29:3284–3292

Lowes R (2017) Heplisav-B vaccine for Hep B finally wins FDA approval. Edited by Medscape

Martinon F, Tschopp J (2005) NLRs join TLRs as innate sensors of pathogens. Trends Immunol 26:447–454

Martins KA, Cooper CL, Stronsky SM, Norris SL, Kwilas SA, Steffens JT, Benko JG, van Tongeren SA, Bavari S (2016) Adjuvant-enhanced CD4 T cell responses are critical to durable vaccine immunity. EBioMedicine 3:67–78

Mata-Haro V, Cekic C, Martin M, Chilton PM, Casella CR, Mitchell TC (2007) The vaccine adjuvant monophosphoryl lipid A as a TRIF-biased agonist of TLR4. Science 316:1628–1632

Matthews K, Chung NP, Klasse PJ, Moutaftsi M, Carter D, Salazar AM, Reed SG, Sanders RW, Moore JP (2013) Clinical adjuvant combinations stimulate potent B-cell responses in vitro by activating dermal dendritic cells. PLoS ONE 8:e63785

McCormack PL, Wagstaff AJ (2006) Ultra-short-course seasonal allergy vaccine (Pollinex Quattro). Drugs 66:931–938

Misquith A, Fung HWM, Dowling QM, Guderian JA, Vedvick TS, Fox CB (2014) In vitro evaluation of TLR4 agonist activity: formulation effects. Colloids Surf, B 113:312–319

Mosca F, Tritto E, Muzzi A, Monaci E, Bagnoli F, Iavarone C, O'Hagan D, Rappuoli R, De Gregorio E (2008) Molecular and cellular signatures of human vaccine adjuvants. Proc Natl Acad Sci U S A 105:10501–10506

Mueller M, Lindner B, Kusumoto S, Fukase K, Schromm AB, Seydel U (2004) Aggregates are the biologically active units of endotoxin. J Biol Chem 279:26307–26313

Mulligan MJ, Bernstein DI, Winokur P et al (2014) Serological responses to an avian influenza a/h7n9 vaccine mixed at the point-of-use with mf59 adjuvant: a randomized clinical trial. JAMA 312:1409–1419

Nguyen M, Leuridan E, Zhang T, De Wit D, Willems F, Van Damme P, Goldman M, Goriely S (2010) Acquisition of adult-like TLR4 and TLR9 responses during the first year of life. PLoS ONE 5:e10407

Nitayaphan S, Khamboonruang C, Sirisophana N, Morgan P, Chiu J, Duliege AM, Chuenchitra C, Supapongse T, Rungruengthanakit K, deSouza M et al (2000) A phase I/II trial of HIV SF2 gp120/MF59 vaccine in seronegative thais. AFRIMS-RIHES Vaccine Evaluation Group. Armed Forces Research Institute of Medical Sciences and the Research Institute for Health Sciences. Vaccine 18:1448–1455

Ohto U, Shibata T, Tanji H, Ishida H, Krayukhina E, Uchiyama S, Miyake K, Shimizu T (2015) Structural basis of CpG and inhibitory DNA recognition by Toll-like receptor 9. Nature 520:702–705

O'Neill LA, Bowie AG (2007) The family of five: TIR-domain-containing adaptors in Toll-like receptor signalling. Nat Rev Immunol 7:353–364

O'Neill LA, Bryant CE, Doyle SL (2009) Therapeutic targeting of Toll-like receptors for infectious and inflammatory diseases and cancer. Pharmacol Rev 61:177–197

Orr MT, Fox CB, Baldwin SL, Sivananthan SJ, Lucas E, Lin S, Phan T, Moon JJ, Vedvick TS, Reed SG et al (2013) Adjuvant formulation structure and composition are critical for the development of an effective vaccine against tuberculosis. J Controlled Release 172:190–200

Otero M, Calarota SA, Felber B, Laddy D, Pavlakis G, Boyer JD, Weiner DB (2004) Resiquimod is a modest adjuvant for HIV-1 gag-based genetic immunization in a mouse model. Vaccine 22:1782–1790

Ozinsky A, Underhill DM, Fontenot JD, Hajjar AM, Smith KD, Wilson CB, Schroeder L, Aderem A (2000) The repertoire for pattern recognition of pathogens by the innate immune system is defined by cooperation between toll-like receptors. Proc Natl Acad Sci U S A 97:13766–13771

Paes W, Brown N, Brzozowski AM, Coler R, Reed S, Carter D, Bland M, Kaye PM, Lacey CJN (2016) Recombinant polymorphic membrane protein D in combination with a novel, second-generation lipid adjuvant protects against intra-vaginal *Chlamydia trachomatis* infection in mice. Vaccine 34:4123–4131

Palm NW, Medzhitov R (2009) Pattern recognition receptors and control of adaptive immunity. Immunol Rev 227:221–233

Panda A, Qian F, Mohanty S, van Duin D, Newman FK, Zhang L, Chen S, Towle V, Belshe RB, Fikrig E et al (2010) Age-associated decrease in TLR function in primary human dendritic cells predicts influenza vaccine response. J Immunol 184:2518–2527

Park BS, Song DH, Kim HM, Choi BS, Lee H, Lee JO (2009) The structural basis of lipopolysaccharide recognition by the TLR4-MD-2 complex. Nature 458:1191–1195

Persing DH, Coler RN, Lacy MJ, Johnson DA, Baldridge JR, Hershberg RM, Reed SG (2002) Taking toll: lipid A mimetics as adjuvants and immunomodulators. Trends Microbiol 10:S32–S37

Phillipps KS, Wykes MN, Liu XQ, Brown M, Blanchfield J, Toth I (2009) A novel synthetic adjuvant enhances dendritic cell function. Immunology 128:e582–e588

Pone EJ, Zhang J, Mai T, White CA, Li G, Sakakura JK, Patel PJ, Al-Qahtani A, Zan H, Xu Z et al (2012) BCR-signalling synergizes with TLR-signalling for induction of AID and immunoglobulin class-switching through the non-canonical NF-κB pathway. Nat Commun 3:767

Prajeeth CK, Jirmo AC, Krishnaswamy JK, Ebensen T, Guzman CA, Weiss S, Constabel H, Schmidt RE, Behrens GM (2010) The synthetic TLR2 agonist BPPcysMPEG leads to efficient cross-priming against co-administered and linked antigens. Eur J Immunol 40:1272–1283

Puggioni F, Durham SR, Francis JN (2005) Monophosphoryl lipid A (MPL) promotes allergen-induced immune deviation in favour of Th1 responses. Allergy 60:678–684

Rallabhandi P, Awomoyi A, Thomas KE, Phalipon A, Fujimoto Y, Fukase K, Kusumoto S, Qureshi N, Sztein MB, Vogel SN (2008) Differential activation of human TLR4 by *Escherichia coli* and *Shigella flexneri* 2a lipopolysaccharide: combined effects of lipid A acylation state and TLR4 polymorphisms on signaling. J Immunol 180:1139–1147

Raman VS, Bhatia A, Picone A, Whittle J, Bailor HR, O'Donnell J, Pattabhi S, Guderian JA, Mohamath R, Duthie MS et al (2010) Applying TLR synergy in immunotherapy: implications in cutaneous leishmaniasis. J Immunol

Ray A, Cot M, Puzo G, Gilleron M, Nigou J (2013) Bacterial cell wall macroamphiphiles: pathogen-/microbe-associated molecular patterns detected by mammalian innate immune system. Biochimie 95:33–42

Reed Steven G, Fox Christopher B, Carter D (2012) Emulsion-based vaccine adjuvants. Future Medicine Ltd.

Rosewich M, Lee D, Zielen S (2013) Pollinex Quattro: an innovative four injections immunotherapy in allergic rhinitis. Hum Vaccin Immunother 9:1523–1531

Ruwona TB, Xu H, Li X, Taylor AN, Shi YC, Cui Z (2016) Toward understanding the mechanism underlying the strong adjuvant activity of aluminum salt nanoparticles. Vaccine 34:3059–3067

Santini-Oliveira M, Coler RN, Parra J, Veloso V, Jayashankar L, Pinto PM, Ciol MA, Bergquist R, Reed SG, Tendler M (2016) Schistosomiasis vaccine candidate Sm14/GLA-SE: Phase 1 safety and immunogenicity clinical trial in healthy, male adults. Vaccine 34:586–594

Schwarz TF (2009) Clinical update of the AS04-Adjuvanted human Papillomavirus-16/18 cervical cancer vaccine, cervarix®. Adv Ther 26:983–998

Seubert A, Monaci E, Pizza M, O'Hagan DT, Wack A (2008) The adjuvants aluminum hydroxide and MF59 induce monocyte and granulocyte chemoattractants and enhance monocyte differentiation toward dendritic cells. J Immunol 180:5402–5412

Seydel U, Oikawa M, Fukase K, Kusumoto S, Brandenburg K (2000) Intrinsic conformation of lipid A is responsible for agonistic and antagonistic activity. Eur J Biochem 267:3032–3039

Shukla NM, Malladi SS, Mutz CA, Balakrishna R, David SA (2010) Structure-activity relationships in human toll-like receptor 7-active imidazoquinoline analogues. J Med Chem 53:4450–4465

Sigal LH, Zahradnik JM, Lavin P, Patella SJ, Bryant G, Haselby R, Hilton E, Kunkel M, Adler-Klein D, Doherty T et al (1998) A vaccine consisting of recombinant *Borrelia burgdorferi* outer-surface protein A to prevent lyme disease. Recombinant outer-surface protein A lyme disease vaccine study consortium. N Engl J Med 339:216–222

Singh M, Khong H, Dai Z, Huang XF, Wargo JA, Cooper ZA, Vasilakos JP, Hwu P, Overwijk WW (2014) Effective innate and adaptive antimelanoma immunity through localized TLR7/8 activation. J Immunol 193:4722–4731

Smirnov D, Schmidt JJ, Capecchi JT, Wightman PD (2011) Vaccine adjuvant activity of 3M-052: an imidazoquinoline designed for local activity without systemic cytokine induction. Vaccine 29:5434–5442

Smith KJ, Hamza S, Skelton H (2003) The imidazoquinolines and their place in the therapy of cutaneous disease. Expert Opin Pharmacother 4:1105–1119

Sogaard OS, Lohse N, Harboe ZB, Offersen R, Bukh AR, Davis HL, Schonheyder HC, Ostergaard L (2010) Improving the immunogenicity of pneumococcal conjugate vaccine in HIV-infected adults with a toll-like receptor 9 agonist adjuvant: a randomized, controlled trial. Clin Infect Dis 51:42–50

Song L, Liu G, Umlauf S, Liu X, Li H, Tian H, Reiserova L, Hou F, Bell R, Tussey L (2014) A rationally designed form of the TLR5 agonist, flagellin, supports superior immunogenicity of Influenza B globular head vaccines. Vaccine 32:4317–4323

Spreafico R, Ricciardi-Castagnoli P, Mortellaro A (2010) The controversial relationship between NLRP3, alum, danger signals and the next-generation adjuvants. Eur J Immunol 40:638–642

Steers NJ, Peachman KK, McClain S, Alving CR, Rao M (2009) Liposome-encapsulated HIV-1 Gag p24 containing lipid A induces effector CD4+ T-cells, memory CD8+ T-cells, and pro-inflammatory cytokines. Vaccine 27:6939–6949

Surquin M, Tielemans C, Nortier J, Jadoul M, Peeters P, Ryba M, Roznovsky L, Doman J, Barthelemy X, Crasta PD et al (2011) Anti-HBs antibody persistence following primary vaccination with an investigational AS02(v)-adjuvanted hepatitis B vaccine in patients with renal insufficiency. Hum vaccines 7:913–918

Takeuchi O, Kawai T, Muhlradt PF, Morr M, Radolf JD, Zychlinsky A, Takeda K, Akira S (2001) Discrimination of bacterial lipoproteins by Toll-like receptor 6. Int Immunol 13:933–940

Tanji H, Ohto U, Shibata T, Taoka M, Yamauchi Y, Isobe T, Miyake K, Shimizu T (2015) Toll-like receptor 8 senses degradation products of single-stranded RNA. Nat Struct Mol Biol 22:109–115

Thoma-Uszynski S, Stenger S, Takeuchi O, Ochoa MT, Engele M, Sieling PA, Barnes PF, Rollinghoff M, Bolcskei PL, Wagner M et al (2001) Induction of direct antimicrobial activity through mammalian toll-like receptors. Science 291:1544–1547

Tighe H, Takabayashi K, Schwartz D, Marsden R, Beck L, Corbeil J, Richman DD, Eiden JJ Jr, Spiegelberg HL, Raz E (2000) Conjugation of protein to immunostimulatory DNA results in a rapid, long-lasting and potent induction of cell-mediated and humoral immunity. Eur J Immunol 30:1939–1947

van Duin D, Shaw AC (2007) Toll-like receptors in older adults. J Am Geriatr Soc 55:1438–1444 van Duin D, Mohanty S, Thomas V, Ginter S, Montgomery RR, Fikrig E, Allore HG, Medzhitov R, Shaw AC (2007) Age-associated defect in human TLR-1/2 function. J Immunol 178:970–975

Van Hoeven N, Fox CB, Granger B, Evers T, Joshi SW, Nana GI, Evans SC, Lin S, Liang H, Liang L et al (2017) A formulated TLR7/8 agonist is a flexible, highly potent and effective adjuvant for pandemic influenza vaccines. Sci Rep 7:46426

Vandepapeliere P, Horsmans Y, Moris P, Van Mechelen M, Janssens M, Koutsoukos M, Van Belle P, Clement F, Hanon E, Wettendorff M et al (2008) Vaccine adjuvant systems containing monophosphoryl lipid A and QS21 induce strong and persistent humoral and T cell responses against hepatitis B surface antigen in healthy adult volunteers. Vaccine 26:1375–1386

Vono M, Taccone M, Caccin P, Gallotta M, Donvito G, Falzoni S, Palmieri E, Pallaoro M, Rappuoli R, Di Virgilio F et al (2013) The adjuvant MF59 induces ATP release from muscle that potentiates response to vaccination. Proc Natl Acad Sci 110:21095–21100

Wen Y, Shi Y (2016) Alum: an old dog with new tricks. Emerg Microbes Infect 5:e25

Wiley SR, Raman VS, Desbien A, Bailor HR, Bhardwaj R, Shakri AR, Reed SG, Chitnis CE, Carter D (2011) Targeting TLRs expands the antibody repertoire in response to a malaria vaccine. Sci Transl Med 3:93ra69

Yoon S-i, Kurnasov O, Natarajan V, Hong M, Gudkov AV, Osterman AL, Wilson IA (2012) Structural basis of TLR5-flagellin recognition and signaling. Science 335:859–864

Zhang X, Mosser DM (2008) Macrophage activation by endogenous danger signals. J Pathol 214:161–178

Zhao BG, Vasilakos JP, Tross D, Smirnov D, Klinman DM (2014) Combination therapy targeting toll like receptors 7, 8 and 9 eliminates large established tumors. J Immunother Cancer 2:12

Zucchini N, Bessou G, Traub S, Robbins SH, Uematsu S, Akira S, Alexopoulou L, Dalod M (2008) Cutting edge: overlapping functions of TLR7 and TLR9 for innate defense against a herpesvirus infection. J Immunol 180:5799–5803

Targeting Glycans on Human Pathogens for Vaccine Design

Stefanie A. Krumm and Katie J. Doores

Contents

Abstract Glycosylation is an important post-translational modification that is required for structural and stability purposes and functional roles such as signalling, attachment and shielding. Many human pathogens such as bacteria display an array of carbohydrates on their surface that are non-self to the host; others such as viruses highjack the host-cell machinery and present self-carbohydrates sometimes arranged in a non-self more immunogenic manner. In combination with carrier proteins, these glycan structures can be highly immunogenic. During natural infection, glycan-binding antibodies are often elicited that correlate with long-lasting protection. A great amount of research has been invested in carbohydrate vaccine

S. A. Krumm · K. J. Doores (✉)
Department of Infectious Diseases, King's College London, 2nd Floor, Borough Wing, Guy's Hospital, London Bridge, London SE1 9RT, UK
e-mail: katie.doores@kcl.ac.uk

S. A. Krumm
e-mail: stefanie.krumm@kcl.ac.uk

Current Topics in Microbiology and Immunology (2020) 428: 129–163
https://doi.org/10.1007/82_2018_103

design to elicit such an immune response, which has led to the development of vaccines against the bacterial pathogens *Haemophilus influenzae type b*, *Streptococcus pneumonia* and *Neisseria meningitidis*. Other vaccines, e.g. against HIV-1, are still in development, but promising progress has been made with the isolation of broadly neutralizing glycan-binding antibodies and the engineering of stable trimeric envelope glycoproteins. Carbohydrate vaccines against other pathogens such as viruses (Dengue, Hepatitis C), parasites (*Plasmodium*) and fungi (*Candida*) are at different stages of development. This chapter will discuss the challenges in inducing cross-reactive carbohydrate-targeting antibodies and progress towards carbohydrate vaccines.

Abbreviations

α-gal	Galα1-3Galβ1-4GlcNAc
Als	Agglutinin-like substance
Asn	Asparagine
bnAbs	Broadly neutralizing antibodies
BSA	Bovine serum albumin
CDR	Complementarity determining region
CRM197	Mutated version of DT
DENV	Dengue virus
DT	Diphteria toxoid
EDE1	E-dimer-dependent epitope 1 group
EDE2	E-dimer-dependent epitope 2 group
Env	Envelope
Gal	Galactose
GalNAc	N-acetylgalactoseamine
GalNAcAN	2-acetamido-2-deoxy-O-D-galact-uronamide
GalXM	Galactoxylomannan
Glc	Glucose
GlcNac	N-acetylglucosamine
GPI	Glycosylphosphatidylinositol
GXM	Glucuronoxylomannan
HepC	Hepatitis C virus
Hib	*Haemophilus influenzae type b*
HIV-1	Human immundeficiency virus type 1
Hwp1	Hyphal wall protein
HYK	Heat-killed yeast
Hyr1	Hyphally regulated protein
KLH	Keyhole limpet haemocyanin
LacDiNAc	GalNAcβ1–4GlcNAc
Lam-CRM197	Laminaran conjugated to CRM197
LDN	GalNAcβ1-4GlcNAc
LDN-F	GalNAcβ1-4(Fucα1-3)GlcNAc
LeX	Galβ1-4(Fucα1-3)GlcNAc

LeX/Lewis X	Galβ1-4(Fucalpha1-3)GlcNAc
Man	Mannose
Men	*Neisseria meningitides*
MurNAc	*N*-acetylmuramic acid
nAbs	Neutralizing antibodies
O-PS	*O*-linked polysaccharide
OMP	Outer-membrane protein
OST	Oligosaccharyltransferase
PCGT	Protein glycan coupling technology
PNGS	Potential glycosylation sites
PRP	Polyribosyl ribitol phosphate
PZQ	Praziquantel
Qui4NFm	4,6-dideoxy-4-formamido-D-glucose
QuiNAc	2-acetamido-2,6-dideoxy-O-D-glucose
Sap	Secreted aspartyl proteases
SEA	Soluble egg antigens
Ser	Serine
SHIV	Simian/human immunodeficiency virus chimera
Sod	Superoxide dismutase
Thr	Threonine
TT	Tetanus toxoid

1 Introduction

The development of vaccines is a great success in the history of medicine, being important in both prevention and eradication of infectious disease in humans. The principle of vaccination was successfully introduced in the late eighteenth century when Edward Jenner used cowpox to immunise humans against smallpox (Riedel 2005). A global vaccination effort, spearheaded by the WHO, led to the eradication of smallpox in 1980 (Fenner 1993). Since then, effective vaccines against diphtheria, tetanus, pertussis, measles, polio, and hepatitis A and B have been developed, which form part of vaccination schedules worldwide (WHO 2015). Although the majority of vaccines bind to protein epitopes within a pathogen, attention is now being turned to targeting glycan structures coating the surface of several major human pathogens. These glycans play roles in structural stability, shielding the pathogen surface from immunodetection, and are important for attachment to host receptors to gain entry into cells.

In 1923, Heidelberger and Avery showed that the sugars coating different *Streptococcus pneumonia* strains were not only diverse in structure but were also recognised by the human immune system and could be used to elicit a long-lasting antibody-mediated protective immune response (Heidelberger and Avery 1923; Heidelberger et al. 1950). Since then, the FDA has licensed carbohydrate-based vaccines in humans against the bacteria *Haemophilus influenzae type b*, *S. pneumonia* and *Neisseria meningitidis*. These vaccines contain pathogen-specific carbohydrates coupled to a carrier protein that induces a glycan-binding antibody response specific for the carbohydrate structures on the pathogen's surface. Although there are currently no licensed carbohydrate-based vaccines against viruses, several antibodies directed against viral glycan structures have been isolated. In particular, glycan-binding broadly neutralizing antibodies (bnAbs) against human immunodeficiency virus 1 (HIV-1) have been shown to protect against SHIV (simian/human immunodeficiency virus chimera) infection in macaque models and highlight the potential for targeting viral glycans for vaccine design (Moldt et al. 2012; Shingai et al. 2014). This chapter will discuss the challenges and prospects for development of carbohydrate-based vaccines against major human pathogens and summarise current research in this area.

2 Glycans on Pathogens as Targets for Vaccine Development

Glycosylation is a post-translational modification, occurring on both proteins (*N*- and *O*-linked glycosylation) and lipids. Many pathogens display carbohydrate structures on their surfaces, and depending on how a pathogen acquires its surface carbohydrates, the host will see these carbohydrates as 'self' or 'non-self'. Pathogens that use their own glycosylation machinery tend to display non-self glycan structures rendering them an attractive epitope for vaccine design. Pathogens that hijack the host cell glycosylation machinery mainly display self-glycans, which poses challenges for the development of vaccines (see Sect. 2.5). However, there are examples where glycosylation can diverge from the host cell leading to self-glycans being displayed in an arrangement that is seen as non-self by the host, e.g. the mannose patch on HIV Envelope. Here, we briefly describe the carbohydrate structures coating the major human pathogens; viruses, bacteria, fungi and parasites (Fig. 1).

2.1 Viruses

Viruses are intracellular parasites that use the host cell to replicate and produce new viral particles to sustain and spread infection. They hijack the host-cell secretion and glycosylation machinery to produce glycoproteins bearing both *N*- and *O*-linked glycans and are therefore mainly recognised as self by the host. *N*-linked glycosylation occurs at Asparagine (Asn) residues within the consensus sequence Asn-X-Threonine (Thr)/Serine (Ser) (where X can be any amino acid except Proline (Pro)). It follows a strict ordered assembly beginning in the ER with addition of $Glc_3Man_9GlcNAc_2$ to Asn within the consensus motif (Fig. 2). This glycan is then trimmed to

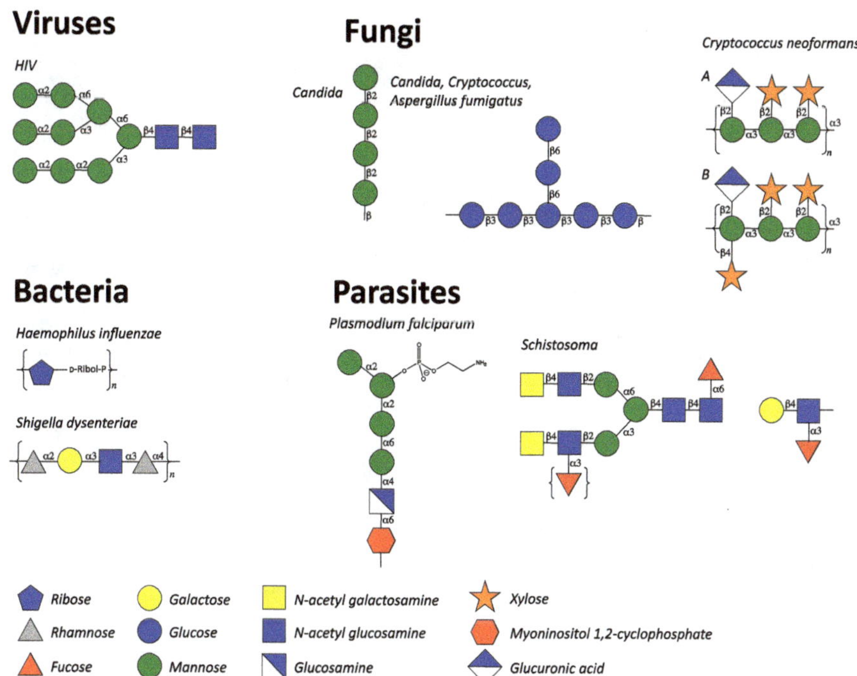

Fig. 1 Representative glycan structures present on bacteria, viruses, parasites and fungi

$Man_5GlcNAc_2$ by the ER- and Golgi-resident mannosidase enzymes. Diversification to complex-type glycans begins with addition of a β1,2-linked GlcNAc residue to $Man_5GlcNAc_2$ in the medial Golgi apparatus. Further trimming and processing in the medial and late Golgi apparatus lead to a wide array of hybrid-type and complex-type glycans and these structures are often dependent on the producer cell type. As this process is not genetically determined, viral glycoproteins tend to consist of a number of glycoforms displaying different glycan structures. Interestingly, HIV-1 displays host-derived high-mannose glycans in clusters. As high-mannose glycans are generally intermediates in the glycosylation pathway and not typically found in mammalian cells, they are recognised as non-self by the immune system (Bonomelli et al. 2011; Doores et al. 2010b) and this cluster forms the epitope of a number of HIV broadly neutralizing antibodies (Figs. 1 and 3) (Walker et al. 2011; Sok et al. 2014; Sanders et al. 2002a; Scanlan et al. 2002). *O*-linked glycosylation occurs on Ser and Thr residues but as there is no precise consensus sequence, it is harder to predict sites of *O*-linked glycosylation. Although most *O*-linked glycosylation occurs in Ser/Thr-rich regions, isolated residues can also be glycosylated. The initial glycan used for attachment to Ser/Thr is predominantly GalNAc and this monosaccharide is then modified by host-cell glycosyltransferases. Viral surface glycoproteins are typically important for cell tropism and infectivity and often hide the antigenic protein from the immune recognition (Vigerust and Shepherd 2007).

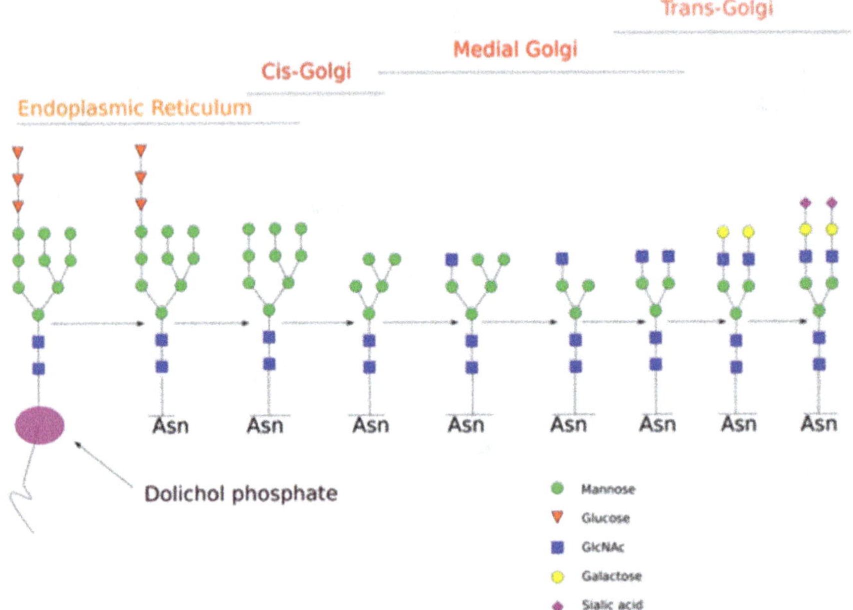

Fig. 2 Schematic of the *N*-linked complex glycan biosynthesis pathway. Glycans are assembled as a $Glc_3Man_9GlcNAc_2$ precursor on the membrane anchor dolichol phosphate on the outside of the ER. The oligosaccharide is flipped inside the ER lumen and transferred onto asparagine residues in the sequon Asn-X-Ser/Thr of the nascent protein chain. Once bound to the protein, glucose is removed and the protein is passed into the Golgi apparatus where the oligosaccharides get further trimmed to $Man_5GlcNAc_2$. Further modifications include the addition of GlcNAc and Gal and/or sialic acid and core fucose to form different variants of complex-type glycans

2.2 Bacteria

Bacteria, both gram-positive and gram-negative, are covered by cross-linked polymer peptidoglycans, which form the major structural component of the periplasm. The peptidoglycan consists of a polysaccharide (composed of *N*-acetylglucosamine and *N*-acetylmuramic acid (MurNAc) in β1-4-linkage) that is covalently cross-linked by short peptides (Brown et al. 2015; Royet and Dziarski 2007). In addition, bacteria produce a polysaccharide capsule that forms the outer surface and mediates adhesion and has a role in virulence (Moxon and Kroll 1990). Antibodies against these polysaccharides are responsible for protection. The structure of the polysaccharide capsule is a unique feature between species and specific serotypes and is determined by the bacterial genome. For example, 90 different polysaccharides have been described for *S. pneumoniae* (Henrichsen 1995). In general, bacterial polysaccharides are made up of a combination of hexuronic acids, *N*-aceylated hexosamines and neutral or amino sugars such as rhamnose. However, some bacteria have polysaccharides containing self-structures, e.g. meningococcus Group B polysaccharide contains sialic acid (Rubens et al. 1987; Finne et al. 1983; Wyle et al. 1972).

2.3   Fungi

The cell wall of fungi consists of complex polysaccharides that are important for maintaining shape and integrity. Fungi possess their own secretory pathway and synthesise their own carbohydrates and glycoconjugates. The fungal wall is highly cross-linked and composed of polymers of galactose, glucose, *N*-acetylglucosamine, mannose, and/or rhamnose, and is specific for the fungal species. The mannose residues in mannans are usually α1–6 linked, whereas the glucose residues in glucans are mostly β1-3 linked (Cummings and Doering 2009). The fungal cell wall also contains glycosylated and/or glycosylphosphatidylinositol (GPI)-anchored proteins and some unusual glycolipids. These sugars are virulence determinants and are highly antigenic.

2.4   Parasites

Parasites pass through multiple developmental life stages during their life cycles often involving different hosts that can impact glycosylation. The glycoconjugates that cover their surface are comprised partly of glycan types typically absent in mammals such as schistosomal Lewis X (LeX, Galβ1-4(Fucalpha1-3)GlcNAc), LacDiNac (LDN, GalNAcβ1-4GlcNAc) or fucosylated LDN (LDN-F, GalNAcβ1-4 (Fucα1-3)GlcNAc) (Fig. 1) and are rich in GPI-anchored proteins, polysaccharides and polysaccharide-binding lectins. These glycoconjugates make up the glycocalyx that is crucial for parasitic virulence (Rodrigues et al. 2015).

2.5   Challenges for the Development of Carbohydrate-Based Vaccines

A number of challenges associated with the generation of antibodies against carbohydrates have meant that the potential of carbohydrate-based vaccines has not yet been fully realised (Scanlan et al. 2007). Firstly, due to the T-cell-independent nature of anti-carbohydrate immune responses, antibody responses against carbohydrates tend to be less robust, short-lived and consist mostly of IgM. Secondly, glycan–protein interactions are inherently weak and therefore binding affinities must be enhanced through multivalent binding and avidity effects. For example, lectins overcome this weak binding by using multiple carbohydrate-binding domains to interact with an array of carbohydrate ligands. Some anti-HIV glycan-binding antibodies bind multiple glycans within one antigen-binding site (Crispin and Doores 2015; Sok et al. 2014; Walker et al. 2011; Sanders et al. 2002a; Scanlan et al. 2002). Thirdly, glycoproteins typically exist as a number of different glycoforms. As protein glycosylation is not directly template driven like protein synthesis, the same protein backbone can be glycosylated with different glycan structures. Further, these glycan structures adopt multiple conformations. These factors increase the microheterogeneity of glycosylation that in turn weakens the antigenic response to carbohydrates. Carbohydrate-binding antibodies need to accommodate this glycan microheterogeneity to avoid further weakening of the antigenic response (Rudd and Dwek 1997). Fourthly, as glycosylation is present in all cells the host may recognise the carbohydrates as self and be tolerated by immune cells.

These combined effects make glycans poorly immunogenic. One of the main concerns and limitations of eliciting antibodies against self-carbohydrates is their potential auto-reactivity and consequent negative selection during B cell development in vivo. A further challenge arises from pathogens that are transmitted between different species. Arthropod-borne pathogens such as Dengue virus (DENV) replicate first in mosquitos (surface proteins dominated by oligo- and paucimannose structures) before they infect humans (proteins dominated by complex-type sugars) (Dejnirattisai et al. 2011; Hacker et al. 2009; Crispin and Doores 2015), and therefore, the carbohydrate epitopes to be targeted by vaccines are different.

2.6 Currently Licensed Carbohydrate-Based Vaccines

Despite the highlighted challenges, several bacterial carbohydrate-based vaccines have been developed that are widely used. In 1977 the first carbohydrate vaccine, PneumoVax (Merck and Co.), was licensed consisting of a mix of 14 unconjugated capsular polysaccharides isolated from 14 different *S. pneumonia* strains. However, this vaccine was not very efficacious in the elderly and immunocompromised and therefore was replaced by the 23-valent vaccine, Pneumovax 23, consisting of 23 polysaccharides identified to be most pathogenic and covering 90% of strains causing serious disease (van Dam et al. 1990).

To overcome the challenges described above, induction of a T cell-dependent immune response against carbohydrates is favourable. To induce a long-lasting memory B cell response, precursor B cells are recruited to and activated in germinal centres formed in secondary lymphoid organs. CD4$^+$ T cells, in particular follicular T-helper cells T_{FH}, assist in affinity maturation, class switching and plasma B cell generation (MacLennan 1994; Sadanand et al. 2016; de Bree and Lynch 2016). The T-cell-independent nature of the carbohydrates can be overcome by chemically conjugating carbohydrates to carrier proteins to recruit CD4$^+$ T cells. Suitable coupling proteins are tetanus toxoid (TT), diphtheria toxoid (DT), a mutated version of DT (CRM197) or the outer-membrane protein (OMP) of *Neisseria meningitidis* (Goldblatt 2000; Ada and Isaacs 2003). These carriers also provide several multiple loci for conjugation to increase multivalency. The first protein-conjugated carbohydrate vaccine was licensed in 1985 and protects against *Haemophilus influenzae type b* (Hib), the most common cause for meningitis and pneumonia. An improved version was approved 2 years later. It contains the main component of the Hib capsule called polyribosyl ribitol phosphate (PRP) conjugated to either tetanus toxoid (PRP-T, Hiberix, Merck an Co.), diphtheria toxoid (Pentacel, Sanofi Pasteur) or the meningococcal OMP (PedvaxHIB, Merk and Co.). In 2000, a conjugated *S. pneumonia* vaccine called PVC 7 was approved (Prevnar, Pfizer). The vaccine contains seven pneumococcal polysaccharides individually conjugated to the diphtheria CRM197 protein. In 2010, six more polysaccharides were added to constitute the PVC 13 vaccine (Prevnar 13, Pfizer). In case of the vaccine against *Neisseria meningitides,* purified capsular polysaccharides from serogroups A, C, W135 and Y (Men ACWY) have been used as vaccines in infants. The meningococcal polysaccharides of Men ACWY consist of homopolymers of either *N*-acetyl mannosamine-1-phosphate in case of serogroup A or sialic acid residues in case of C, W135 and Y and are poorly immunogenic. However, chemical coupling of the polysaccharides to DT (Menactra, Sanofi Pasteur) or to CRM97 (Menveo, Novartis Vaccines and Diagnostics) enhances immunogenicity greatly (US Food and Drug Administration 2015). Vaccines against other glycosylated pathogens are under development and will be discussed in the following sections.

3 Vaccines Under Development Against Major Human Pathogens

3.1 Viruses

Thus far, no carbohydrate-based vaccine against viral glycans has been developed. Unlike some other pathogens, enveloped viruses use the host-cell glycosylation machinery to attach host-derived glycans to their viral surface and are therefore typically thought to be poorly immunogenic. However, recent advances in the isolation of antibodies from infected patients have led to the isolation of neutralizing antibodies (nAbs) that bind to carbohydrates on viral surface glycoproteins.

3.1.1 HIV-1 Envelope Glycans Are a Target for HIV-1 Broadly Neutralizing Antibodies

The main example of glycan-binding anti-viral neutralizing antibodies is HIV-1 broadly neutralizing antibodies (bnAbs). These are effective against a large number of circulating strains within each virus group and are able to prevent infection when passively administered to macaques in challenge studies (Moldt et al. 2012). Characterisation of their epitopes and their longitudinal development in vivo can give valuable information for vaccine design. HIV-1 infects immune cells (CD4$^+$ T cells, dendritic cells and macrophages) and integrates into the host genome. There is currently no cure or vaccine and anti-retroviral drugs are the only effective treatment.

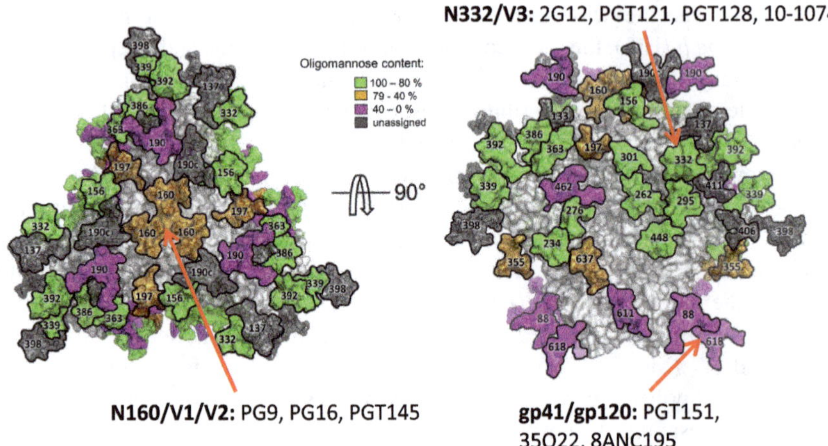

Fig. 3 Location of *N*-linked glycans on HIV-1 Env important for glycan-binding HIV-1 bnAbs. Model shows location of glycans and the processing state. Red arrows indicate bnAbs targeting highlighted glycan sites (Reproduced and adapted from Behrens et al. Cell Reports 2016)

HIV-1 Envelope (Env) is the sole target for HIV bnAbs. Env consists of a trimer of gp41/gp120 heterodimers. It is first expressed as the precursor gp160, which is glycosylated in the ER and further processed in the Golgi and cleaved via Furin into gp120 and gp41. HIV Env is highly glycosylated with up to 50% of its mass consisting of host-derived *N*-linked glycans making it one of the most heavily glycosylated proteins known. In each monomeric gp120, there are approximately 25–30 potential glycosylations sites (PNGS) (Coss et al. 2016). It is thought that these glycans are important for transmission through interaction with lectins such as DC-SIGN but the glycans also shield conserved regions of the protein from the immune system. Therefore, the glycans on Env are often referred to as the glycan shield or silent face. Although the glycans are attached via the host-cell glycosylation machinery, analysis of the glycans present on Env shows an unusually high abundance of oligomannose-type glycans (Fig. 1). It has been shown that the dense clustering of *N*-linked glycosylation sites restricts access by the ER α-mannosidase I enzyme, thus preventing cleavage of $Man_{8–9}GlcNAc_2$ glycans resulting in a patch of underprocessed non-self high-mannose glycans (Bonomelli et al. 2011; Crispin and Doores 2015; Doores et al. 2010a; Go et al. 2013; Zhou et al. 2002; Behrens et al. 2016). This cluster is often referred to as the intrinsic mannose patch and is conserved across geographical HIV clades. The additional restrictions associated with trimerization results in a further elevation in oligomannose glycans leading to the trimer-associated mannose patch (Bonomelli et al. 2011; Crispin and Doores 2015; Doores et al. 2010a; Alexandre et al. 2010; Eggink et al. 2010).

Despite the challenges associated with eliciting glycan-binding antibodies described above, some HIV-infected individuals, after 2–3 years of infection, elicit a neutralizing antibody response that is targeted against the glycan structures on

HIV Env and can neutralise up to 80% of circulating HIV strains (Goo et al. 2012; Sok et al. 2014). These bnAbs target three distinct glycan regions on Env. The most potent bnAb families target an area, which centres on glycan N332 on the outer domain of gp120 and their epitopes include protein residues in the V3 loop in addition to glycans at N332/N334, N301 and/or N295 (e.g. PGT121, PGT128, 10-1074) (Fig. 3) (Walker et al. 2011; Kong et al. 2013; Mouquet et al. 2012). The bnAb 2G12 targets the same region but only binds to glycans. 2G12 uses a unique domain-exchange structure to form a multivalent binding surface that enhances binding through increased avidity (Calarese et al. 2003). The other two glycan epitopes centre, firstly, on N160 and protein residues in the V1/V2 loop (PG9, PG16, PGT145) (Julien et al. 2013b; Pancera et al. 2013) and, secondly, on glycans at the interface of gp120 and gp41, including N88, N611 and N637 and surrounding protein residues (e.g. PGT151, 35O22, 8ANC195) (Falkowska et al. 2014; Huang et al. 2014; Scharf et al. 2015) (Fig. 3). Generally, these HIV bnAbs overcome the weak glycan–protein interactions by binding to multiple glycans and protein residues within one Fab.

Passive administration of HIV bnAbs to macaques is able to protect from SHIV infection in challenge studies (Moldt et al. 2012). These studies suggest that if these bnAbs could be elicited through vaccination then they would be able to prevent infection in humans (Moldt et al. 2012; Shingai et al. 2014). However, there are a number of challenges associated with eliciting such bnAbs through vaccination. They are highly mutated (30–40% compared to 10–15% found in other neutralizing antibodies) and often contain multiple insertions and deletions as well as a long complementarity determining regions H3 (CDRH3) that penetrates through the glycan shield to reach the protein surface beneath (e.g. PG9, PGT128) (Scheid et al. 2009; Walker et al. 2011; McLellan et al. 2011). Further, it can take 2–3 years of chronic infection before bnAbs can be detected (Kwong et al. 2013). Neutralisation escape usually occurs through addition or deletion of N-linked glycan sites. However, several bnAbs can accommodate viral escape by binding promiscuously to other glycans in close proximity (Doores et al. 2015; Krumm et al. 2016; Sok et al. 2014).

3.1.2   Vaccine Strategies for Eliciting Glycan-Binding HIV bnAbs

Research in the field of HIV vaccines is increasingly highlighting the HIV glycan shield as a good target for HIV vaccine design. BnAbs targeting the glycans on HIV-1 Env are highly desirable due to their high potency and breadth, and their ability to bind alternative arrangements of glycans within the mannose patch. Several strategies have been and are currently being investigated.

Initial design strategies focused on chemical synthesis of clusters of high-mannose glycans. These consisted of precise oligodendron scaffolds or multivalent arrays of glycans on viral particles such as bacteriophage Qβ (Astronomo et al. 2010; Doores et al. 2010b). Although these immunogens were able to elicit mannose-binding antibodies, they were unable to neutralise HIV-1. Similarly, using antigenic mimicry of yeast carbohydrates to induce 2G12 like antibodies was also unsuccessful (Agrawal-Gamse et al. 2011; Wang 2013). Immunization with monomeric gp120 or gp140 has been attempted in a number of studies but non-neutralising antibodies were mainly produced (Flynn et al. 2005; Pitisuttithum et al. 2006). A major breakthrough occurred with the development of a stabilised native-like trimeric Env (BG505 SOSIP.664). The trimer was stabilised through the introduction of a disulphide bond to covalently link the gp41 and gp120 subunits, and the addition of trimer-stabilizing mutations between gp41 subunits (Julien et al. 2013a; Sanders et al. 2002b, 2013; Khayat et al. 2013; Binley et al. 2000). The quaternary structure of this protein was shown to be important for correct glycosylation of Env (Pritchard et al. 2015). Immunization of SOSIP trimers derived from a viral strain with moderate sensitivity to neutralizing antibody (classified as tier 2) (Seaman et al. 2010) in small and non-human primate animal models produced homologous tier 2 neutralizing antibodies; however, heterologous neutralizing antibodies against other tier 2 viruses was not detected (Sanders et al. 2015). It is not known whether the elicited antibodies directly contact glycan structures on Env.

Current immunogen design strategies are now focused on recapitulating the bnAb affinity maturation process occurring in HIV-infected individuals. This involves characterising the bnAb lineage from germline and determining Env sequences that drive development of this lineage (Mouquet 2015; Jardine et al. 2013; Yang et al. 2015). It is thought that stepwise administration of rationally designed immunogens to activate B cell precursors might drive antibody maturation. Although current strategies are focused on eliciting CD4 binding site bnAbs (Jardine et al. 2015; Dosenovic et al. 2015; Jardine et al. 2016), these approaches have also been explored for glycan-binding HIV bnAbs (Steichen et al. 2016).

3.1.3 Targeting Glycans on Other Viral Pathogens for Vaccine Design

Although not as well studied as HIV, there are other viral targets for which glycan-binding bnAbs have been isolated.

Hepatitis C

Hepatitis C (HCV) is a major human pathogen with more than 170 million people suffering from chronic HCV infection leading to liver cirrhosis or liver cancer. Although progress has been made in the development of treatment and a cure, these are very expensive and a vaccine is desperately needed. The HCV glycoprotein consists of a heterodimer of E1/E2 and is the target of neutralizing antibodies. E1 and E2 have up to 6 and 11 PNGS, respectively, and these glycans are predominantly high mannose due to budding of viral particles from the ER.

A number of bnAbs have been isolated that potently inhibit HCV infection in vitro (Srinivasan et al. 2016). Antibodies AR3A-D target discontinuous protein and glycan residues within E2 (regions 396–424, 436–447 and 523–554) with Ser424, Gly523, Pro525, Gly530, Asp535, Val538 and Asn540 constituting conserved contact residues preventing receptor engagement (Law et al. 2008). AR4A is a potent neutralizing antibody but only recognises protein residues (Drummer and Poumbourios 2004). When the three most potent antibodies AR3A, AR3B and AR4A were generated via adeno-associated virus vectors in mice, all three antibodies potently inhibited HCV infection in vitro (de Jong et al. 2014; Law et al. 2008). Humanised mice that express the three AAV-derived antibodies were protected from infection when challenged with HCV. Furthermore, AR3A, AR3B and AR4A can abrogate an established HCV infection in vitro in primary human foetal liver cultures (de Jong et al. 2014). The therapeutic effect of these antibodies has also been demonstrated in vivo. When passively administered to mice with an established HCV infection, a single dose of the bnAbs reduced the genome copy number to detection limit. Essentially, these animals can be regarded as cured, probably by protecting healthy liver cells from infection and allowing for clearance of HCV in already infected hepatocytes (de Jong et al. 2014). Passive immunization using these bnAbs could be used as a feasible therapeutic, particularly in patients unresponsive to other anti-viral therapies, although antibody doses need to be determined in humans. Vaccine development targeting the conserved AR regions with immunogen-induced antibodies is a promising strategy.

Dengue Virus

Dengue virus (DENV), a member of the Flaviviridae family, is a mosquito-borne virus infecting about 400 million people yearly, predominantly in South America and the Pacific Islands, and there is currently no treatment. Neutralizing antibodies have been identified from infected individuals that target the surface glycoprotein, E. At the initial infection, the glycan profile of the surface protein is dominated by high- or paucimannose N-glycans consistent with production in insect cells while after the virus replicates in the human host, the glycans switch to complex type (Dejnirattisai et al. 2011; Hacker et al. 2009). Antibodies 747(4) A11, 747 B7, 752 (2) C8 and 753(3) C10 contact the highly conserved glycans at positions N67 and N153. Antibodies A11 and B7 belong to the E-dimer-dependent epitope 2 group (EDE2) and are strictly dependent on the glycan at position N153, while C8 and C10 belong to the E-dimer-dependent epitope 1 group (EDE1) and can bind without N153 being glycosylated. The different glycan dependencies arise due to different mechanisms of binding to the protein component of the epitope. EDE2 and EDE1 both bind to the valley formed by the β strand on the domain II side (residues 67–74, 97–106 and 246–249). EDE2 interacts with the '150 loop' of domain I (residues 148–159) containing N153 while EDE1 targets domains I and III and induces disorder of the 150 loop (Rouvinski et al. 2015). In a similar way to HIV, high levels of somatic mutation have been shown to be important for binding. Current vaccine design strategies use a similar approach as for HIV-1 vaccines by using a stabilised pre-fusion E dimer to stimulate B cells that give rise to DENV bnAbs.

Recently, a recombinant, live, attenuated, tetravalent dengue vaccine has just been approved in Brazil, the Philippines and Mexico only. Dengvaxia (Sanofi Pasteur) is safe and effective against serotypes 1–4 yet it is not known if this vaccine elicits antibodies targeting the *N*-linked glycans on E.

Arenaviruses

Finally, and in contrast to the viruses described above, arenaviruses do not appear to induce any effective glycan-binding neutralizing antibody response (Sommerstein et al. 2015). Lassa virus, the most prominent member of the Old World arenavirus family, is endemic in West Africa causing haemorrhagic fever in severe cases and ultimately leading to death in up to 10% of the cases. Other members are of the New World arenaviruses causing Argentine, Venezuelan, Bolivian and Brazilian haemorrhagic fever are Junin, Guanarito, Machupo and Sabia viruses, respectively (Charrel and de Lamballerie 2003). Glycans present on the surface proteins GP do not prevent specific antibody induction per se but counteract neutralization by reducing the antibody on-rate and occupancy and prevent antibody-mediated virus control. This may explain why antibody immunity induced by either vaccination or the natural infection tends to be low. In this case, carbohydrate vaccines would have the opposite effects and are therefore not a good target for vaccine design against arenaviruses (Sommerstein et al. 2015).

3.2 *Bacteria*

As discussed above, several conjugate and polysaccharide-based vaccines have been developed against bacterial pathogens. However, there are many clinically relevant bacteria for which vaccine development is still ongoing. Bacterial vaccines are particularly challenging due to the heterogeneity and complexity of capsular polysaccharides (Astronomo and Burton 2010). Further, bacterial glycans can be very similar to human glycans, which can result in immune tolerance and poor immunogenicity. Bacterial glycans can be chemically modified to enhance immunogenicity. However, these modification needs to be finely balanced to maintain the glycan integrity for the immune system to still be able to recognise the pathogen but be distinguishable from the human version (Astronomo and Burton 2010). Current glycoconjugate vaccines are produced by purification of the polysaccharide from the native pathogen followed by chemical coupling to a carrier protein. This process can lead to impurities and batch-to-batch variation, and therefore, novel approaches using synthetic carbohydrates are currently being investigated. Synthetic vaccines allow reproducible production of well-defined conjugates free from contaminants. This system can be adapted for use with many serotypes and multiple serotype combinations are possible. Synthetic carbohydrate vaccines offer the possibility of varying the sugar chain length and coupling

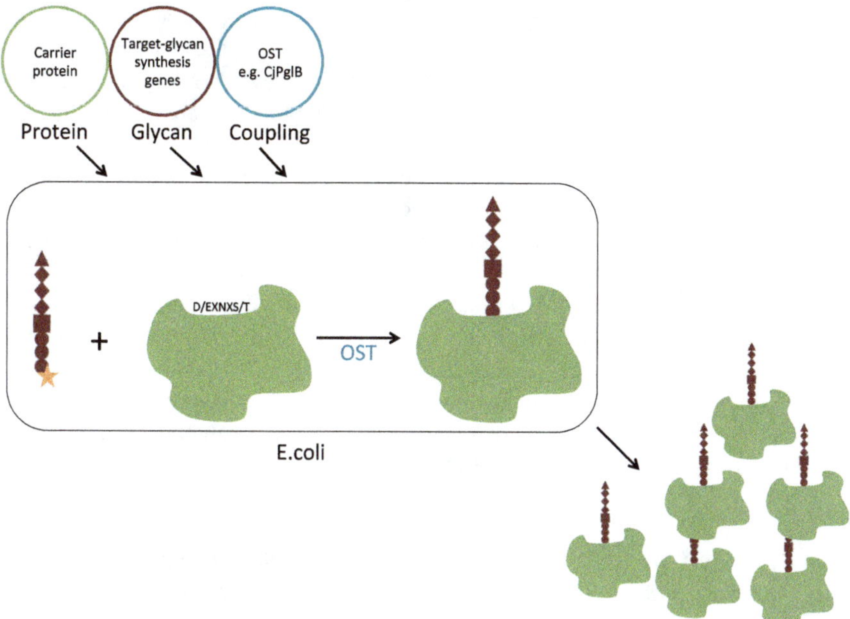

Fig. 4 Schematic of the protein glycan coupling technology. *E. coli* cells are transformed with three bacterial expression plasmids encoding for the carrier protein harbouring the glycan acceptor sequon D/EXNXS/T, the polysaccharide locus and the oligosaccharyltransferase (OST) PglB. The OST transfers the glycan containing the reducing end sugar (orange star) onto the carrier protein to produce the desired glycoprotein in large amounts (adapted from Cuccui et al. 2013 and Terra et al. 2012)

density, as well as chemical modification of the glycan, to induce the optimal antibody response (Pozsgay 2008). Overall, synthetic carbohydrate vaccines help to understand the relationship between chain length, saccharide composition and secondary structured by creating a flexible platform to improve immunogenicity (Astronomo and Burton 2010).

3.2.1 *Shigella*

Shigella causes approximately 1.1 million deaths each year with 60% being children under 5 years of age (Kotloff et al. 1999). There are approximately 50 serotypes grouped into 5 serogroups. *S. sonnei* is the most prevalent in developed countries, while *S. flexneri* is prevalent in developing countries. The *O*-linked polysaccharides (O-PS) are a major component in the bacterial cell wall and determine the serotype. To protect against the major circulating strains, a multivalent vaccine needs to be developed. A vaccine containing *S. dysenteriae* type 1, *S. sonnei*, *S. flexneri* 2a, 3, and 6 would protect against 75% of *Shigella* outbreaks (Levine et al. 2007). The inclusion of the three *S. flexneri* O-PS would cross-protect against all 11 serotypes as they share the same polysaccharide antigen (Noriega et al. 1999). Conventional conjugation of *Shigella* *O*-polysaccharide covalently linked to a carrier protein has given good efficacy, however not in young children (Passwell et al. 2010). Therefore, new vaccine approaches are being pursued. Current research is focused on synthetic production of more immunogenic conjugate vaccines. For example, a synthetic pentasaccharide made of three repeating units of the O-PS of *S. flexneri* 2a is recognised by antibodies in serum from infected individuals and can convey protection in mice (Phalipon et al. 2009). When administered as a conjugate linked to TT in the presence of the adjuvant alum, it induced an increased anti-*S flexneri* 2a antibody response that can be sustained for more than one year (van der Put et al. 2016). Furthermore, potential vaccine candidates against *S. dysenteriae* include multiple repeats of the tetrasaccharide of the O-PS. Different densities of tetra-, octa-, dodeca- and hexadecasaccharides were tested, but only the octa-, dodeca- and hexadeca-conjugates were able to induce an immune response (Pozsgay et al. 1999).

3.2.2 Protein-Glycan Coupling Technology to Develop Vaccines Against *Shigella Dysenteriae* and *Francisella Tularensis*

An experimental vaccine candidate produced with a new technique called protein glycan coupling technology (PGCT) is under investigation against *Shigella dysenteriae* and *Francisella tularensis* (Fig. 4). PGCT can be applied to numerous combinations of recombinant protein–glycan structures and replaces in vitro conjugation of polysaccharide to proteins. This method uses a coupling enzyme oligosaccharyltransferase (OST, for example, PglB isolated from *Campylobacter jejuni*) that transfers the reducing end of a sugar to a consensus recognition sequence in the carrier protein (D/EYNXS/T). All three components are cloned into bacterial expression plasmids and transformed into the appropriate *E. coli* strain to express a recombinant glycan–protein conjugate that can easily be produced in large scale and purified in a single step (Terra et al. 2012). Additional advantages are the maintenance of structural integrity due to a quick turnover and the usage of *E. coli* over more pathogenic bacteria. The carrier proteins might not contain the glycosylation consensus sequon; however, they can be engineered onto the target protein (Fisher et al. 2011). The limiting factor for this method is the substrate specificity of the OST PglB. It only transfers glycans with a reducing end sugar containing an acetamido group in the C2 position (Wacker et al. 2006). Other PglBs with broader specificity are under investigation (Terra et al. 2012). A glycan-conjugate vaccine against *S. dysenteria* using PCGT has been developed by GlycoVaxyn AG. The O-PS was coupled to the carrier protein exotoxin A of *Pseudomonas aeruginosa*. A Phase I trail demonstrated the vaccine candidate was not only safe but elicited a good antibody response against *Shigella* O-PS at a range of doses (Hatz et al. 2015).

The bacterium *F. tularensis*, a class A bioterrorism agent, causes tularaemia and is very infectious at low doses, has a high fatality rate and is easily transmitted via aerosol. Attempts to use a live vaccine failed as it caused disease in the mouse model (Fortier et al. 1991). Purification of the O-antigen from the pathogen [a tetrasaccharide with the structure 4-α-D-GalNAcAN-(1–4)-α-D-GalNAcAN-(1–3)-β-D-QuiNAc-(1–2)-β-D-Qui4NFm-(1-)] is challenging due to the difficulties associated with handling of highly infectious agents. Therefore in the current strategy, the O-antigen is produced in *E. coli* and then coupled to carrier protein exotoxin A from *P. aeruginosa* with the help of the OST PglB (Prior et al. 2003). This glycoconjugate vaccine showed promising efficacy in a mouse model, demonstrating that the PGCT method might be a valuable strategy to produce glycoconjugate vaccines against challenging to work with bacterial pathogens (Cuccui et al. 2013).

3.3 Fungi

Fungi can cause severe infections, particularly in hospital settings when a patient's immune system is compromised. Despite there being thousands of fungal species, only a select few cause human disease including several species of *Candida* as well as other mycoses like *Aspergillum* and *Cryptococcus*, and dimorphic fungi such as *Histoplasma capsulatum* or *Blastomyces dermatitidis*. The protective cell wall of fungi is generally covered by antigenic polysaccharides such as β-1,3-glucan, β-1,6-glucan, α-mannan, β-mannan and chitin (a linear homopolymer of β-1,4-linked N-acetylglucosamine). Together, they form the scaffold for predominantly GP1-anchored glycoproteins that are themselves extensively modified with *N*-linked and *O*-linked oligosaccharides (Klis et al. 2011; Plaine et al. 2008; Bowman and Free 2006). The major obstacle for broad vaccine development is the divergence in cell wall structure between fungal strains. Most vaccine strategies have been univalent consisting of one antigen rendering them strain specific. High treatment costs highlight the need for an effective vaccine in both healthy and immunocompromised patients that induces long-lasting immunological memory. It is thought that this can be achieved by combining mechanisms of innate and adaptive immune response (Roy and Klein 2012; Dockrell et al. 1999; Leung et al. 2004; Levin et al. 2001; Madhi et al. 2005; Nordoy et al. 2002; Tedaldi et al. 2004).

3.3.1 Candida Albicans

The most prevalent fungal infection worldwide is caused by *Candida albicans*, a component of the commensal gut microbiota. This polymorphic organism exists in two life forms: unicellular (yeast) or multicellular (hyphae) and it is the hyphal form that causes the disease. Breaches in the tissue barrier (most common cause, triggered mainly by surgery, catheters, fasting, extended hospital stay) altered gut flora (through use of broad-spectrum antibiotics) or immune suppression can lead to the development of mucosal candidiasis. In the most severe cases, the fungus circulates in the bloodstream causing a highly lethal candidemia and can infect organs (Cassone 2013; Diekema et al. 2012; Pappas 2006).

C. albicans displays a wide variety of antigenic factors such as members of the heavily glycosylated agglutinin-like substance (Als), superoxide dismutase (Sod) or secreted aspartyl proteases (Sap) families (Cassone 2013). Thus far, two vaccines targeting protein antigens have successfully completed Phase I clinical trials. The first consists of rAls3p-N (recombinant N-terminus of Als3p, NDV-3, NovaDigm Therapeutics) (Schmidt et al. 2012) and the second (PEV-7, Pevion Biotech) relies on amino acids 77–400 of the secreted Sap2p (De Bernardis et al. 2012, 2015). The Als3-containing vaccine includes both a T_H1 and T_H17 cell-based immune response as well as an Als3-specific antibodies response. Even though these approaches induce a promising immune response, they fail to induce a persistent protection due to univalency of the strategy (Cassone 2013). Glycoconjugate immunogens could be a promising alternative for a multivalent vaccine strategy.

Glycoconjugate vaccines consisting of β-glucan and β-mannan polysaccharides conjugated to tetanus or diphtheria toxoid (Torosantucci et al. 2005; Xin et al. 2008) have been shown to elicit the required T cell response as well as induce antibodies that cross-protect against major fungal pathogens including *Asperillus* spp. and *Cryptococcus* spp. (Gaffen et al. 2011; Kaufmann 2007; Romani 2011; Spellberg et al. 2008). These vaccines could be delivered via virosomes or other nanoparticles. Those carriers have been used successfully in other vaccine candidates (e.g. against influenza) and been shown to induce a robust T cell response (Cusi 2006; Moser et al. 2013; Radosevic et al. 2008; Zhao et al. 2014). Immunization with the β-1,3-glucan polysaccharide laminaran, isolated from algae and conjugated to CRM197 (Lam-CRM197), in combination with the adjuvant MF59 was found to reduce mortality when mice were challenged with a lethal dose of *C. albicans* (Bromuro et al. 2010; Pietrella et al. 2010). This vaccine was also effective against *Aspergillus*, and *Cryptococcus* suggesting cross-protection against fungi is possible by targeting the glucans in their cell wall (Bromuro et al. 2010; Torosantucci et al. 2005).

Since the fungus is also part of the commensal microbiota, the ideal vaccine would induce antibodies specific for the disease causing hyphal form of *Candida* (Cassone 2013). Attenuated or inactivated full organism versions, while able to induce long-lasting immunity for viral or bacterial pathogens (Minor 2015), have to be different enough to prevent cross-reactivity with commensal yeast strains (Saville et al. 2009). Therefore, a more specific approach being investigated focuses on coupling the polysaccharides to hyphal associated proteins to increase specific anti-fungal efficacy. Two such proteins, expressed as virulence factors in the cell wall during the yeast-to-hypha transition, are the hyphally regulated protein (Hyr1) or the hyphal wall protein (Hwp1) (Cassone et al. 2010; Luo et al. 2010; Xin et al. 2008). A β-mannan trisaccharide coupled to Hwp1 [β-(Man)3-Hwp1]-induced significant protection (80–100%) in mice studies (Xin et al. 2008), rendering this glycoconjugate vaccine approach highly promising. Of note, such antibodies could also be administered safely to treat fungal infections passively in infants and immunocompromised patients.

3.3.2 *Aspergillus Fumigatus*

The second most common fungal infection is caused by *Aspergillus fumigatus* leading to allergic bronchopulmonary or invasive aspergillosis and other fungal diseases and has a very high mortality rate particularly in immunocompromised individuals. Vaccination against *A. fumigatus* with crude antigen preparations has been shown to be effective in immunocompromised mice (Ito and Lyons 2002). The immunodominant antigen in the crude extract has been determined: Asp f3, a peroxisomal protein in the hyphae (Diaz-Arevalo et al. 2011), and when the protein was administered subcutaneously, it protected mice from a lethal dose of *A. fumigatus* (Ito et al. 2006). With regard to glycans, galactofuranose-containing glycolipids (glycosylinositolphosphoceramid and GPI-anchored lipophospho-galactomannan) as well as polysaccharide (α1-3glucan, β1–3glucan, and galactomannan) induced a robust immune response in mice, activating IL-17 or IFN-γ, IL-17-, and IL-10-producing T cells, respectively (Bozza et al. 2009). Antibodies against β1–3glucan partially contribute to the immune response (Torosantucci et al. 2005).

3.3.3 *Cryptococcus Neoformans*

Among the *Crytococcus* spp., *Cryptococcus neoformans* is the third most common cause of invasive fungal diseases. The capsule of *C. neoformans* consists predominantly of the polysaccharides α-1,3-D-mannopyranose units with single residues of ß-D-xylopyranosyl and ß-D-glucuronopyranosyl attached called glucuronoxylomannan (GXM) and two minor components, galactoxylomannan (GalXM) and mannoprotein rendering it the prime target for vaccine design (Cherniak and Sundstrom 1994). This dominant virulence factor GXM causes an immunodysregulatory effect and thereby suppresses the host inflammatory response and prevents opsonophagocytosis (Shoham et al. 2001; Spellberg 2011). However, when GXM coupled to a carrier such as TT is administered to mice, it induces a protective antibody response (Mukherjee et al. 1993). The same is true for GalXM when the polysaccharide its conjugated to bovine serum albumin (BSA) or a protective antigen of Bacillus anthracis (Chow and Casadevall 2011). Even passively administered GXM-specific antibodies protect mice from cryptococcosis (Casadevall et al. 1998; Mukherjee et al. 1993). However, GXM vaccines also elicit non-specific or deleterious antibodies that bind to other sites in GXM reducing the protective efficacy of the vaccine (Devi et al. 1991; Mukherjee et al. 1995). Additionally, GXM exerts immunomodulatory effects that interfere with leukocyte migration (Ellerbroek et al. 2004; Vecchiarelli 2000). To assess the side effects of these deleterious antibodies in humans, therapeutic efficacy of passively administered antibodies was assessed in a phase I dose-escalation study with the goal to establish the safety of antibody administration in patients with cryptococcal antigenemia. HIV-1 infected patients with a history of culture-proven and treated monococcal meningitis were administered with dfferent doses of murineanticryptococcal monoclonal antibody 18B7 and the study found that the antibody is safe to be administered at doses up to 1.0 mg/kg without evidence of toxicity as suggested by in vitro or animal studies (Larsen et al. 2005).

Alternative approaches under investigation include a peptide mimotope (P13) of the polysaccharide GXM conjugated to either BSA or tetanus toxoid. These vaccines were shown to prolong survival of mice infected with a lethal dose of *C. neoformans* (Fleuridor et al. 2001). As mentioned above, the laminaran (β-1,3-glucan polysaccharide) vaccine also cross-protects against *C. neoformans*. Other examples of protective antigens constituting the capsule include the less abundant mannoproteins. Yeast proteins contain generally terminal mannosylated glycans, which are distinct from host glycans. The role of terminal mannosylation, more the *O*- mannosylation than the *N*- mannosylation, has been found to be crucial for mediating a potent T cell-mediated immune response, $CD8^+$-T cell function and proliferation as well as secretion of pro-inflammatory cytokines such as TNF and IL-12 (Specht et al. 2007; Lam et al. 2005; Luong et al. 2007) and should be taken into consideration for inclusion in future vaccine design strategies to stimulate the immune system additionally and hence increase efficacy of any other vaccine (Levitz et al. 2015).

Multivalent, broad-spectrum anti-fungal vaccines are the ultimate goal to achieve. A multivalent vaccine would provide broad, additive and synergistic protection reducing the chances of immune evasion as has been achieved for *S. pneumonia* vaccination. Heat-killed *Saccharomyces cerevisiae* (heat-killed yeast, HKY) vaccination has been shown to protect mice against systemic aspergilliosis (Liu et al. 2011), coccidioidomycosis (Capilla et al. 2009), candidiasis (Liu et al. 2012) and cryptococcosis (Majumder et al. 2014) suggesting a broad vaccine could be possible if the right immunogens were identified. Furthermore, a combination of common cell wall polysaccharide components, e.g. β-glucan and β-mannan or mannoproteins, conjugated to fungal proteins, such as Candida proteins Als3, Hwp1 or Hyr1 alone or in combination, maybe a promising approach to enhance not only efficacy but also breadth of the immune response (Casadevall and Pirofski 2006; Johnson and Bundle 2013). However, efficacy, cross-reactivity, and sustainability of the immune response need to be determined in animal models, as well as in humans.

3.4 Parasites

Parasites are complex organisms both genetically and biologically and cause diseases such as schistosomiasis, malaria, toxoplasmosis and leishmaniasis, and are caused by protozoan parasites. Several peptide-, protein- and DNA-based vaccines have been tested, but with limited success (Graves and Gelband 2003). Glycan and peptide antigens are very complex, not always present on all parasite life stages and hence are relatively poorly understood. It is only recently that glycans have been identified as being a dominant antigen on certain parasites (Hokke and Deelder

2001; Thomas and Harn 2004). However, vaccine development has been slow as it is difficult to obtain enough glycan material due to challenges in culturing of parasites or complex synthetic protocols. This section focuses on the trematode *Schistosome mansoni* and the protozoan parasite *Plasmodium falciparum*.

3.4.1 Schistosomes

Schistosomiasis is caused by trematodes (helminthic parasites) such as *Schistosoma mansoni, S. intercalatum, S. haematobium, S. japonicum* and *S. mekongi.* The life cycle involves both human and snail hosts and the different life stages (larval (cercaria), schistosomula, adult worms and eggs) present different glycan structures that also vary in their abundance. It is the egg stage that typically causes severe disease. Eggs spread disease through shedding into the urine or faeces. When the eggs get trapped in the human tissue they cause fibrosis, calcification, hepatosplenomegaly and potentially fatal bleeding from oesophageal varices (Richter et al. 1998). During schistosome infection, a robust antibody response is induced, however this response is not very protective and immunity can take many years to develop (Eberl et al. 2001b; Hagan and Sharaf 2003). Praziquantel (PZQ) is the only approved drug for treatment of schistosome infection. Many parasite immunogenic proteins have been investigated in vaccine studies, but with limited efficacy (Hewitson and Maizels 2014; Hagan and Sharaf 2003). Glycans rather than proteins are the immunodominant antigens with antibodies mainly targeting the glycan structures on parasitic glycoproteins (Nyame et al. 2004). The development of glycan microarrays displaying parasite glycans has allowed the identification of antigenic motifs (van Diepen et al. 2012a, 2015). One such array is a shotgun microarray where glycans isolated directly from the parasite are printed on glass slides allowing unique and unusual parasite-specific glycans to be included that are not available through chemical synthesis (van Diepen et al. 2012b, 2015; de Boer et al. 2008). Antibodies elicited during infection target specific *N*- and *O*-linked antigenic structures containing the motifs LacDiNAc (LDN, GalNAcβ1-4GlcNAc), LDN-F (GalNAcβ1-4 (Fucα1-3)GlcNAc), Lewis X (LeX, Galβ1-4(Fucα1-3)GlcNAc) and core α3-fucose/ β2-xylose (Fig. 1). These motifs are present in all parasite life stages but vary in surface expression levels and exposure. Terminal β-linked GalNAc is an example of a non-self parasitic motif not found in vertebra. LDN is continuously present while LeX and LDN-F are mainly synthesised after transformation into schistosomula (Smit et al. 2015). Importantly, LeX targeting IgM, IgG as well as IgA have been isolated, while LDN and LDN-F targeting IgG and core α3-fucose/β2-xylose targeting IgE have been found (Van Roon et al. 2004).

A vaccine producing antibodies against glycans and glycoproteins present on the different parasite life stages is likely required for full protection in humans. A vaccine consisting of soluble egg antigens (SEA) conjugated to cholera toxin B subunit has been trailed in mice and shown to induce IgE, enhance the Th2 response, and reduce liver granuloma and mortality (Sun et al. 2001; Okano et al. 1999). Radiation-attenuated schistosomes have been successfully used to induce a

IgM and IgG response (Delgado and McLaren 1990; Hewitson et al. 2005; Eberl et al. 2001a). The protective antigen has not been identified yet and the application of these vaccine approaches in humans is questionable. Parasite glycans are particularly of interest for vaccine design due to their high density and multivalent display on the parasite surface. However, it is imperative to identify the correct glycan structures to achieve the most specific and potent immune response. This is an area still under investigation (Luyai et al. 2014; Mandalasi et al. 2013; Prasanphanich et al. 2014) and the development of glycan microarrays will strongly inform these studies.

3.4.2 *Plasmodium*

Another very well-studied protozoan parasite of the genus *Plasmodium* causes Malaria in humans. *P. falciparum* is the most pathogenic and is transmitted as sporozoites via mosquitos. It infects hepatocytes in the liver where it matures to merozoites which can then infect red blood cells. The parasite causes red blood cell lysis and release, and it spreads to other red blood cells causing severe tissue pathology. Current vaccine strategies target several different stages of the parasite life cycle, pre-erythrocytic, erythrocytic and transmission blocking. The malaria vaccine Mosquirix^{TM} (RTS,S by GSK) is the most advanced pre-erythrocytic vaccine candidate and contains portions of the circumsporozoite protein. It is about 30% effective against the most prevalent parasite in Africa, *P. falciparum*, but is not effective against *P. vivax*, which is prevalent outside of Africa. The requirement of four doses further complicates the application of this vaccine (Todryk and Hill 2007).

Novel vaccine strategies are focused on GPI anchors, which are dominant toxins responsible for the pathology. They are important in signal transduction processes by activating macrophages and vascular endothelial cells leading to production of pro-inflammatory mediators such as nitrous oxide, tumour necrosis factor α and intercellular adhesion molecule-1 (Schofield et al. 1996; Tachado et al. 1996; Nyame et al. 2004). GPIs are produced synthetically and contain multiple mannose residues, glucosamine and 6-myoinositol-1,2-cyclic phosphate (NH_2-CH_2-CH_2-PO_4-(Manα1-2)6Manα1-2Manα1-6Manα1-4GlcNH$_2\alpha$1-6*myo*-inositol-1,2-cyclic phosphate) conjugated to the carrier keyhole limpet haemocyanin (KLH) (Fig. 1) (Hewitt et al. 2002; Liu et al. 2005; Schofield et al. 2002). In vivo studies showed good IgG titre induction against parasitic GPI, but importantly not mammalian GPI. Immunization of mice with the GPI conjugate resulted in a 75% reduction in death rate after *Plasmodium* challenge and showed cross-reactivity with other *Plasmodium* species, also to those only infecting mice. This phenotype demonstrates a conserved GPI structure. However, the GPI conjugate vaccine did not prevent infection per se nor did it cause parasite death but was found instead to neutralise toxicity associated with infection (Nyame et al. 2004).

The carbohydrate epitope Galα1-3Galβ1-4GlcNAc-R (α-gal) is a structure that was deleted from the human glycan repertoire during evolution. α-gal is therefore a xeno-antigen and is expressed by bacterial components of the microbiota. 1–5% of circulating IgG and IgM in humans recognises this glycan structure (Macher and Galili 2008). Interestingly, *Plasmodium* sporozoites have α-gal on the surface, whether α-gal is produced by the parasite itself of by the mosquito is not yet clear (Galili et al. 1998; Warburg et al. 2007). It has been shown that α-gal antibodies elicited against the gut microbiota protect from *P. falciparum* infections in humans and transmissions in mice experiments (Yilmaz et al. 2014). However, during natural infection in humans, α-gal antibodies levels are too low to be protective. Strategies to boost this promising α-gal immune response, e.g. via addition of adjuvants or by coupling the α-gal to *Plasmodium*-specific antigens, are under investigation as a vaccine candidate (Benatuil et al. 2005; Yilmaz et al. 2014). This strategy could potentially protect against other vector-borne protozoan parasites expressing α-gal, such as *Leishmania* spp. and *Trypanosoma* spp.

4 Discussion

There is a lot of activity in the field of carbohydrate-based vaccine research with some promising candidates in the pipeline. Although there are currently only carbohydrate-based vaccines against bacterial pathogens, the potential for their development against other pathogens is high. This is particularly true for viral pathogens considering the isolation of many glycan-binding neutralizing antibodies. However, the approaches used are quite different. For viral pathogens, a reverse vaccinology methodology is being utilised where functional antibodies are isolated from natural infection and then characterisation of their binding epitopes is informing immunogen design in an attempt to re-elicit these antibodies in vivo. For other glycosylated pathogens the approach is to generate immunogens that contain surface glycans/polysaccharides coupled to protein carriers. With the development of high-throughput glycan microarray methodologies and antibody isolation techniques, it would be possible to use this reverse vaccinology approach with the other glycosylated pathogens described above. In this way, additional information on how glycan-binding antibodies generated from natural infection differ from those generated from vaccination and may inform vaccine design against many different relevant pathogens.

References

Ada G, Isaacs D (2003) Carbohydrate-protein conjugate vaccines. Clin Microbiol Infect 9(2):79–85
Agrawal-Gamse C, Luallen RJ, Liu B, Fu H, Lee FH, Geng Y, Doms RW (2011) Yeast-elicited cross-reactive antibodies to HIV Env glycans efficiently neutralize virions expressing exclusively high-mannose N-linked glycans. J Virol 85(1):470–480. https://doi.org/10.1128/JVI.01349-10

Alexandre KB, Gray ES, Lambson BE, Moore PL, Choge IA, Mlisana K, Karim SS, McMahon J, O'Keefe B, Chikwamba R, Morris L (2010) Mannose-rich glycosylation patterns on HIV-1 subtype C gp120 and sensitivity to the lectins, Griffithsin Cyanovirin-N and Scytovirin. Virology 402(1):187–196. https://doi.org/10.1016/j.virol.2010.03.021

Astronomo RD, Burton DR (2010) Carbohydrate vaccines: developing sweet solutions to sticky situations? Nat Rev Drug Discov 9(4):308–324. https://doi.org/10.1038/nrd3012

Astronomo RD, Kaltgrad E, Udit AK, Wang SK, Doores KJ, Huang CY, Pantophlet R, Paulson JC, Wong CH, Finn MG, Burton DR (2010) Defining criteria for oligomannose immunogens for HIV using icosahedral virus capsid scaffolds. Chem Biol 17(4):357–370. https://doi.org/10.1016/j.chembiol.2010.03.012

Behrens AJ, Vasiljevic S, Pritchard LK, Harvey DJ, Andev RS, Krumm SA, Struwe WB, Cupo A, Kumar A, Zitzmann N, Seabright GE, Kramer HB, Spencer DI, Royle L, Lee JH, Klasse PJ, Burton DR, Wilson IA, Ward AB, Sanders RW, Moore JP, Doores KJ, Crispin M (2016) Composition and antigenic effects of individual glycan sites of a trimeric HIV-1 envelope glycoprotein. Cell Rep 14(11):2695–2706. https://doi.org/10.1016/j.celrep.2016.02.058

Benatuil L, Kaye J, Rich RF, Fishman JA, Green WR, Iacomini J (2005) The influence of natural antibody specificity on antigen immunogenicity. Eur J Immunol 35(9):2638–2647. https://doi.org/10.1002/eji.200526146

Binley JM, Sanders RW, Clas B, Schuelke N, Master A, Guo Y, Kajumo F, Anselma DJ, Maddon PJ, Olson WC, Moore JP (2000) A recombinant human immunodeficiency virus type 1 envelope glycoprotein complex stabilized by an intermolecular disulfide bond between the gp120 and gp41 subunits is an antigenic mimic of the trimeric virion-associated structure. J Virol 74(2):627–643

Bonomelli C, Doores KJ, Dunlop DC, Thaney V, Dwek RA, Burton DR, Crispin M, Scanlan CN (2011) The glycan shield of HIV is predominantly oligomannose independently of production system or viral clade. PLoS ONE 6(8):e23521. https://doi.org/10.1371/journal.pone.0023521

Bowman SM, Free SJ (2006) The structure and synthesis of the fungal cell wall. BioEssays 28 (8):799–808. https://doi.org/10.1002/bies.20441

Bozza S, Clavaud C, Giovannini G, Fontaine T, Beauvais A, Sarfati J, D'Angelo C, Perruccio K, Bonifazi P, Zagarella S, Moretti S, Bistoni F, Latge JP, Romani L (2009) Immune sensing of *Aspergillus fumigatus* proteins, glycolipids, and polysaccharides and the impact on Th immunity and vaccination. J Immunol 183(4):2407–2414. https://doi.org/10.4049/jimmunol. 0900961

Bromuro C, Romano M, Chiani P, Berti F, Tontini M, Proietti D, Mori E, Torosantucci A, Costantino P, Rappuoli R, Cassone A (2010) Beta-glucan-CRM197 conjugates as candidates antifungal vaccines. Vaccine 28(14):2615–2623. https://doi.org/10.1016/j.vaccine.2010.01.012

Brown L, Wolf JM, Prados-Rosales R, Casadevall A (2015) Through the wall: extracellular vesicles in Gram-positive bacteria, mycobacteria and fungi. Nat Rev Microbiol 13(10):620–630. https://doi.org/10.1038/nrmicro3480

Calarese DA, Scanlan CN, Zwick MB, Deechongkit S, Mimura Y, Kunert R, Zhu P, Wormald MR, Stanfield RL, Roux KH, Kelly JW, Rudd PM, Dwek RA, Katinger H, Burton DR, Wilson IA (2003) Antibody domain exchange is an immunological solution to carbohydrate cluster recognition. Science 300(5628):2065–2071. https://doi.org/10.1126/science.1083182

Capilla J, Clemons KV, Liu M, Levine HB, Stevens DA (2009) *Saccharomyces cerevisiae* as a vaccine against coccidioidomycosis. Vaccine 27(27):3662–3668. https://doi.org/10.1016/j. vaccine.2009.03.030

Casadevall A, Pirofski LA (2006) Polysaccharide-containing conjugate vaccines for fungal diseases. Trends Mol Med 12(1):6–9. https://doi.org/10.1016/j.molmed.2005.11.003

Casadevall A, Cleare W, Feldmesser M, Glatman-Freedman A, Goldman DL, Kozel TR, Lendvai N, Mukherjee J, Pirofski LA, Rivera J, Rosas AL, Scharff MD, Valadon P, Westin K, Zhong Z (1998) Characterization of a murine monoclonal antibody to *Cryptococcus neoformans* polysaccharide that is a candidate for human therapeutic studies. Antimicrob Agents Chemother 42(6):1437–1446

Cassone A (2013) Development of vaccines for *Candida albicans*: fighting a skilled transformer. Nat Rev Microbiol 11(12):884–891. https://doi.org/10.1038/nrmicro3156

Cassone A, Bromuro C, Chiani P, Torosantucci A (2010) Hyr1 protein and beta-glucan conjugates as anti-Candida vaccines. J Infect Dis 202(12):1930. https://doi.org/10.1086/657417

Charrel RN, de Lamballerie X (2003) Arenaviruses other than Lassa virus. Antiviral Res 57(1–2):89–100

Cherniak R, Sundstrom JB (1994) Polysaccharide antigens of the capsule of *Cryptococcus neoformans*. Infect Immun 62(5):1507–1512

Chow SK, Casadevall A (2011) Evaluation of *Cryptococcus neoformans* galactoxylomannan-protein conjugate as vaccine candidate against murine cryptococcosis. Vaccine 29(10):1891–1898. https://doi.org/10.1016/j.vaccine.2010.12.134

Coss KP, Vasiljevic S, Pritchard LK, Krumm SA, Glaze M, Madzorera S, Moore PL, Crispin M, Doores KJ (2016) HIV-1 glycan density drives the persistence of the mannose patch within an infected individual. J Virol 90(24):11132–11144. https://doi.org/10.1128/JVI.01542-16

Crispin M, Doores KJ (2015) Targeting host-derived glycans on enveloped viruses for antibody-based vaccine design. Curr Opin Virol 11:63–69. https://doi.org/10.1016/j.coviro.2015.02.002

Cuccui J, Thomas RM, Moule MG, D'Elia RV, Laws TR, Mills DC, Williamson D, Atkins TP, Prior JL, Wren BW (2013) Exploitation of bacterial N-linked glycosylation to develop a novel recombinant glycoconjugate vaccine against *Francisella tularensis*. Open Biol 3(5):130002. https://doi.org/10.1098/rsob.130002

Cummings RD, Doering TL (2009) Fungi. In: Essentials of glycobiology, vol 2. Cold Spring Harbor Laboratory Press

Cusi MG (2006) Applications of influenza virosomes as a delivery system. Hum Vaccin 2(1):1–7

De Bernardis F, Amacker M, Arancia S, Sandini S, Gremion C, Zurbriggen R, Moser C, Cassone A (2012) A virosomal vaccine against candidal vaginitis: immunogenicity, efficacy and safety profile in animal models. Vaccine 30(30):4490–4498. https://doi.org/10.1016/j.vaccine.2012.04.069

De Bernardis F, Arancia S, Sandini S, Graziani S, Norelli S (2015) Studies of immune responses in candida vaginitis. Pathogens 4(4):697–707. https://doi.org/10.3390/pathogens4040697

de Boer AR, Hokke CH, Deelder AM, Wuhrer M (2008) Serum antibody screening by surface plasmon resonance using a natural glycan microarray. Glycoconj J 25(1):75–84. https://doi.org/10.1007/s10719-007-9100-x

de Bree GJ, Lynch RM (2016) B cells in HIV pathogenesis. Curr Opin Infect Dis 29(1):23–30. https://doi.org/10.1097/QCO.0000000000000225

de Jong YP, Dorner M, Mommersteeg MC, Xiao JW, Balazs AB, Robbins JB, Winer BY, Gerges S, Vega K, Labitt RN, Donovan BM, Giang E, Krishnan A, Chiriboga L, Charlton MR, Burton DR, Baltimore D, Law M, Rice CM, Ploss A (2014) Broadly neutralizing antibodies abrogate established hepatitis C virus infection. Sci Transl Med 6(254):254ra129. https://doi.org/10.1126/scitranslmed.3009512

Dejnirattisai W, Webb AI, Chan V, Jumnainsong A, Davidson A, Mongkolsapaya J, Screaton G (2011) Lectin switching during dengue virus infection. J Infect Dis 203(12):1775–1783. https://doi.org/10.1093/infdis/jir173

Delgado V, McLaren DJ (1990) Evidence for enhancement of IgG1 subclass expression in mice polyvaccinated with radiation-attenuated cercariae of *Schistosoma mansoni* and the role of this isotype in serum-transferred immunity. Parasite Immunol 12(1):15–32

Devi SJ, Schneerson R, Egan W, Ulrich TJ, Bryla D, Robbins JB, Bennett JE (1991) *Cryptococcus neoformans* serotype A glucuronoxylomannan-protein conjugate vaccines: synthesis, characterization, and immunogenicity. Infect Immun 59(10):3700–3707

Diaz-Arevalo D, Bagramyan K, Hong TB, Ito JI, Kalkum M (2011) CD4[+] T cells mediate the protective effect of the recombinant Asp f3-based anti-aspergillosis vaccine. Infect Immun 79(6):2257–2266. https://doi.org/10.1128/IAI.01311-10

Diekema D, Arbefeville S, Boyken L, Kroeger J, Pfaller M (2012) The changing epidemiology of healthcare-associated candidemia over three decades. Diagn Microbiol Infect Dis 73(1):45–48. https://doi.org/10.1016/j.diagmicrobio.2012.02.001

Dockrell DH, Poland GA, Steckelberg JM, Wollan PC, Strickland SR, Pomeroy C (1999) Immunogenicity of three *Haemophilus influenzae type b* protein conjugate vaccines in HIV seropositive adults and analysis of predictors of vaccine response. Vaccine 17(22):2779–2785

Doores KJ, Bonomelli C, Harvey DJ, Vasiljevic S, Dwek RA, Burton DR, Crispin M, Scanlan CN (2010a) Envelope glycans of immunodeficiency virions are almost entirely oligomannose antigens. Proc Natl Acad Sci U S A 107(31):13800–13805. https://doi.org/10.1073/pnas. 1006498107

Doores KJ, Fulton Z, Hong V, Patel MK, Scanlan CN, Wormald MR, Finn MG, Burton DR, Wilson IA, Davis BG (2010b) A nonself sugar mimic of the HIV glycan shield shows enhanced antigenicity. Proc Natl Acad Sci U S A 107(40):17107–17112. https://doi.org/10. 1073/pnas.1002717107

Doores KJ, Kong L, Krumm SA, Le KM, Sok D, Laserson U, Garces F, Poignard P, Wilson IA, Burton DR (2015) Two classes of broadly neutralizing antibodies within a single lineage directed to the high-mannose patch of HIV envelope. J Virol 89(2):1105–1118. https://doi.org/ 10.1128/JVI.02905-14

Dosenovic P, von Boehmer L, Escolano A, Jardine J, Freund NT, Gitlin AD, McGuire AT, Kulp DW, Oliveira T, Scharf L, Pietzsch J, Gray MD, Cupo A, van Gils MJ, Yao KH, Liu C, Gazumyan A, Seaman MS, Bjorkman PJ, Sanders RW, Moore JP, Stamatatos L, Schief WR, Nussenzweig MC (2015) Immunization for HIV-1 broadly neutralizing antibodies in human Ig knockin mice. Cell 161(7):1505–1515. https://doi.org/10.1016/j.cell.2015.06.003

Drummer HE, Poumbourios P (2004) Hepatitis C virus glycoprotein E2 contains a membrane-proximal heptad repeat sequence that is essential for E1E2 glycoprotein heterodimerization and viral entry. J Biol Chem 279(29):30066–30072. https://doi.org/10. 1074/jbc.M405098200

Eberl M, Langermans JA, Frost PA, Vervenne RA, van Dam GJ, Deelder AM, Thomas AW, Coulson PS, Wilson RA (2001a) Cellular and humoral immune responses and protection against schistosomes induced by a radiation-attenuated vaccine in chimpanzees. Infect Immun 69(9):5352–5362

Eberl M, Langermans JA, Vervenne RA, Nyame AK, Cummings RD, Thomas AW, Coulson PS, Wilson RA (2001b) Antibodies to glycans dominate the host response to schistosome larvae and eggs: is their role protective or subversive? J Infect Dis 183(8):1238–1247. https://doi.org/ 10.1086/319691

Eggink D, Melchers M, Wuhrer M, van Montfort T, Dey AK, Naaijkens BA, David KB, Le Douce V, Deelder AM, Kang K, Olson WC, Berkhout B, Hokke CH, Moore JP, Sanders RW (2010) Lack of complex N-glycans on HIV-1 envelope glycoproteins preserves protein conformation and entry function. Virology 401(2):236–247. https://doi.org/10.1016/j.virol. 2010.02.019

Ellerbroek PM, Walenkamp AM, Hoepelman AI, Coenjaerts FE (2004) Effects of the capsular polysaccharides of *Cryptococcus neoformans* on phagocyte migration and inflammatory mediators. Curr Med Chem 11(2):253–266

Falkowska E, Le KM, Ramos A, Doores KJ, Lee JH, Blattner C, Ramirez A, Derking R, van Gils MJ, Liang CH, McBride R, von Bredow B, Shivatare SS, Wu CY, Chan-Hui PY, Liu Y, Feizi T, Zwick MB, Koff WC, Seaman MS, Swiderek K, Moore JP, Evans D, Paulson JC, Wong CH, Ward AB, Wilson IA, Sanders RW, Poignard P, Burton DR (2014) Broadly neutralizing HIV antibodies define a glycan-dependent epitope on the prefusion conformation of gp41 on cleaved envelope trimers. Immunity 40(5):657–668. https://doi.org/10.1016/j. immuni.2014.04.009

Fenner F (1993) Smallpox: emergence, global spread, and eradication. Hist Philos Life Sci 15 (3):397–420

Finne J, Leinonen M, Makela PH (1983) Antigenic similarities between brain components and bacteria causing meningitis. Implications for vaccine development and pathogenesis. Lancet 2 (8346):355–357

Fisher AC, Haitjema CH, Guarino C, Celik E, Endicott CE, Reading CA, Merritt JH, Ptak AC, Zhang S, DeLisa MP (2011) Production of secretory and extracellular N-linked glycoproteins in *Escherichia coli*. Appl Environ Microbiol 77(3):871–881. https://doi.org/10.1128/AEM. 01901-10

Fleuridor R, Lees A, Pirofski L (2001) A cryptococcal capsular polysaccharide mimotope prolongs the survival of mice with *Cryptococcus neoformans* infection. J Immunol 166(2):1087–1096

Flynn NM, Forthal DN, Harro CD, Judson FN, Mayer KH, Para MF, rgp HIVVSG (2005) Placebo-controlled phase 3 trial of a recombinant glycoprotein 120 vaccine to prevent HIV-1 infection. J Infect Dis 191 (5):654–665. https://doi.org/10.1086/428404

Fortier AH, Slayter MV, Ziemba R, Meltzer MS, Nacy CA (1991) Live vaccine strain of *Francisella tularensis*: infection and immunity in mice. Infect Immun 59(9):2922–2928

Gaffen SL, Hernandez-Santos N, Peterson AC (2011) IL-17 signaling in host defense against *Candida albicans*. Immunol Res 50(2–3):181–187. https://doi.org/10.1007/s12026-011-8226-x

Galili U, LaTemple DC, Radic MZ (1998) A sensitive assay for measuring alpha-Gal epitope expression on cells by a monoclonal anti-Gal antibody. Transplantation 65(8):1129–1132

Go EP, Liao HX, Alam SM, Hua D, Haynes BF, Desaire H (2013) Characterization of host-cell line specific glycosylation profiles of early transmitted/founder HIV-1 gp120 envelope proteins. J Proteome Res 12(3):1223–1234. https://doi.org/10.1021/pr300870t

Goldblatt D (2000) Conjugate vaccines. Clin Exp Immunol 119(1):1–3

Goo L, Jalalian-Lechak Z, Richardson BA, Overbaugh J (2012) A combination of broadly neutralizing HIV-1 monoclonal antibodies targeting distinct epitopes effectively neutralizes variants found in early infection. J Virol 86(19):10857–10861. https://doi.org/10.1128/JVI. 01414-12

Graves P, Gelband H (2003) Vaccines for preventing malaria. Cochrane Database Syst Rev (1): CD000129. https://doi.org/10.1002/14651858.cd000129

Hacker K, White L, de Silva AM (2009) N-linked glycans on dengue viruses grown in mammalian and insect cells. J Gen Virol 90(Pt 9):2097–2106. https://doi.org/10.1099/vir.0.012120-0

Hagan P, Sharaf O (2003) Schistosomiasis vaccines. Expert Opin Biol Ther 3(8):1271–1278. https://doi.org/10.1517/14712598.3.8.1271

Hatz CF, Bally B, Rohrer S, Steffen R, Kramme S, Siegrist CA, Wacker M, Alaimo C, Fonck VG (2015) Safety and immunogenicity of a candidate bioconjugate vaccine against *Shigella dysenteriae* type 1 administered to healthy adults: a single blind, partially randomized Phase I study. Vaccine 33(36):4594–4601. https://doi.org/10.1016/j.vaccine.2015.06.102

Heidelberger M, Avery OT (1923) The soluble specific substance of pneumococcus. J Exp Med 38 (1):73–79

Heidelberger M, Dilapi MM, Siegel M, Walter AW (1950) Persistence of antibodies in human subjects injected with pneumococcal polysaccharides. J Immunol 65(5):535–541

Henrichsen J (1995) Six newly recognized types of *Streptococcus pneumoniae*. J Clin Microbiol 33(10):2759–2762

Hewitson JP, Maizels RM (2014) Vaccination against helminth parasite infections. Expert Rev Vaccines 13(4):473–487. https://doi.org/10.1586/14760584.2014.893195

Hewitson JP, Hamblin PA, Mountford AP (2005) Immunity induced by the radiation-attenuated schistosome vaccine. Parasite Immunol 27(7–8):271–280. https://doi.org/10.1111/j.1365-3024. 2005.00764.x

Hewitt MC, Snyder DA, Seeberger PH (2002) Rapid synthesis of a glycosylphosphatidylinositol-based malaria vaccine using automated solid-phase oligosaccharide synthesis. J Am Chem Soc 124(45):13434–13436

Hokke CH, Deelder AM (2001) Schistosome glycoconjugates in host-parasite interplay. Glycoconj J 18(8):573–587

Huang J, Kang BH, Pancera M, Lee JH, Tong T, Feng Y, Imamichi H, Georgiev IS, Chuang GY, Druz A, Doria-Rose NA, Laub L, Sliepen K, van Gils MJ, de la Pena AT, Derking R,

Klasse PJ, Migueles SA, Bailer RT, Alam M, Pugach P, Haynes BF, Wyatt RT, Sanders RW, Binley JM, Ward AB, Mascola JR, Kwong PD, Connors M (2014) Broad and potent HIV-1 neutralization by a human antibody that binds the gp41-gp120 interface. Nature 515 (7525):138–142. https://doi.org/10.1038/nature13601

Ito JI, Lyons JM (2002) Vaccination of corticosteroid immunosuppressed mice against invasive pulmonary aspergillosis. J Infect Dis 186(6):869–871. https://doi.org/10.1086/342509

Ito JI, Lyons JM, Hong TB, Tamae D, Liu YK, Wilczynski SP, Kalkum M (2006) Vaccinations with recombinant variants of *Aspergillus fumigatus* allergen Asp f 3 protect mice against invasive aspergillosis. Infect Immun 74(9):5075–5084. https://doi.org/10.1128/IAI.00815-06

Jardine J, Julien JP, Menis S, Ota T, Kalyuzhniy O, McGuire A, Sok D, Huang PS, MacPherson S, Jones M, Nieusma T, Mathison J, Baker D, Ward AB, Burton DR, Stamatatos L, Nemazee D, Wilson IA, Schief WR (2013) Rational HIV immunogen design to target specific germline B cell receptors. Science 340(6133):711–716. https://doi.org/10.1126/science.1234150

Jardine JG, Ota T, Sok D, Pauthner M, Kulp DW, Kalyuzhniy O, Skog PD, Thinnes TC, Bhullar D, Briney B, Menis S, Jones M, Kubitz M, Spencer S, Adachi Y, Burton DR, Schief WR, Nemazee D (2015) HIV-1 VACCINES. Priming a broadly neutralizing antibody response to HIV-1 using a germline-targeting immunogen. Science 349(6244):156–161. https://doi.org/10.1126/science.aac5894

Jardine JG, Kulp DW, Havenar-Daughton C, Sarkar A, Briney B, Sok D, Sesterhenn F, Ereno-Orbea J, Kalyuzhniy O, Deresa I, Hu X, Spencer S, Jones M, Georgeson E, Adachi Y, Kubitz M, deCamp AC, Julien JP, Wilson IA, Burton DR, Crotty S, Schief WR (2016) HIV-1 broadly neutralizing antibody precursor B cells revealed by germline-targeting immunogen. Science 351(6280):1458–1463. https://doi.org/10.1126/science.aad9195

Johnson MA, Bundle DR (2013) Designing a new antifungal glycoconjugate vaccine. Chem Soc Rev 42(10):4327–4344. https://doi.org/10.1039/c2cs35382b

Julien JP, Cupo A, Sok D, Stanfield RL, Lyumkis D, Deller MC, Klasse PJ, Burton DR, Sanders RW, Moore JP, Ward AB, Wilson IA (2013a) Crystal structure of a soluble cleaved HIV-1 envelope trimer. Science 342(6165):1477–1483. https://doi.org/10.1126/science. 1245625

Julien JP, Lee JH, Cupo A, Murin CD, Derking R, Hoffenberg S, Caulfield MJ, King CR, Marozsan AJ, Klasse PJ, Sanders RW, Moore JP, Wilson IA, Ward AB (2013b) Asymmetric recognition of the HIV-1 trimer by broadly neutralizing antibody PG9. Proc Natl Acad Sci U S A 110(11):4351–4356. https://doi.org/10.1073/pnas.1217537110

Kaufmann SH (2007) The contribution of immunology to the rational design of novel antibacterial vaccines. Nat Rev Microbiol 5(7):491–504. https://doi.org/10.1038/nrmicro1688

Khayat R, Lee JH, Julien JP, Cupo A, Klasse PJ, Sanders RW, Moore JP, Wilson IA, Ward AB (2013) Structural characterization of cleaved, soluble HIV-1 envelope glycoprotein trimers. J Virol 87(17):9865–9872. https://doi.org/10.1128/JVI.01222-13

Klis FM, de Koster CG, Brul S (2011) A mass spectrometric view of the fungal wall proteome. Future Microbiol 6(8):941–951. https://doi.org/10.2217/FMB.11.72

Kong L, Lee JH, Doores KJ, Murin CD, Julien JP, McBride R, Liu Y, Marozsan A, Cupo A, Klasse PJ, Hoffenberg S, Caulfield M, King CR, Hua Y, Le KM, Khayat R, Deller MC, Clayton T, Tien H, Feizi T, Sanders RW, Paulson JC, Moore JP, Stanfield RL, Burton DR, Ward AB, Wilson IA (2013) Supersite of immune vulnerability on the glycosylated face of HIV-1 envelope glycoprotein gp120. Nat Struct Mol Biol 20(7):796–803. https://doi.org/10. 1038/nsmb.2594

Kotloff KL, Winickoff JP, Ivanoff B, Clemens JD, Swerdlow DL, Sansonetti PJ, Adak GK, Levine MM (1999) Global burden of Shigella infections: implications for vaccine development and implementation of control strategies. Bull World Health Organ 77(8):651–666

Krumm SA, Mohammed H, Le KM, Crispin M, Wrin T, Poignard P, Burton DR, Doores KJ (2016) Mechanisms of escape from the PGT128 family of anti-HIV broadly neutralizing antibodies. Retrovirology 13(1):8. https://doi.org/10.1186/s12977-016-0241-5

Kwong PD, Mascola JR, Nabel GJ (2013) Broadly neutralizing antibodies and the search for an HIV-1 vaccine: the end of the beginning. Nat Rev Immunol 13(9):693–701. https://doi.org/10.1038/nri3516

Lam JS, Mansour MK, Specht CA, Levitz SM (2005) A model vaccine exploiting fungal mannosylation to increase antigen immunogenicity. J Immunol 175(11):7496–7503

Larsen RA, Pappas PG, Perfect J, Aberg JA, Casadevall A, Cloud GA, James R, Filler S, Dismukes WE (2005) Phase I evaluation of the safety and pharmacokinetics of murine-derived anticryptococcal antibody 18B7 in subjects with treated *Cryptococcal meningitis*. Antimicrob Agents Chemother 49(3):952–958. https://doi.org/10.1128/AAC.49.3.952-958.2005

Law M, Maruyama T, Lewis J, Giang E, Tarr AW, Stamataki Z, Gastaminza P, Chisari FV, Jones IM, Fox RI, Ball JK, McKeating JA, Kneteman NM, Burton DR (2008) Broadly neutralizing antibodies protect against hepatitis C virus quasispecies challenge. Nat Med 14 (1):25–27. https://doi.org/10.1038/nm1698

Leung TF, Li CK, Hung EC, Chan PK, Mo CW, Wong RP, Chik KW (2004) Immunogenicity of a two-dose regime of varicella vaccine in children with cancers. Eur J Haematol 72(5):353–357. https://doi.org/10.1111/j.1600-0609.2004.00216.x

Levin MJ, Gershon AA, Weinberg A, Blanchard S, Nowak B, Palumbo P, Chan CY, Team ACTG (2001) Immunization of HIV-infected children with varicella vaccine. J Pediatr 139(2):305–310. https://doi.org/10.1067/mpd.2001.115972

Levine MM, Kotloff KL, Barry EM, Pasetti MF, Sztein MB (2007) Clinical trials of Shigella vaccines: two steps forward and one step back on a long, hard road. Nat Rev Microbiol 5 (7):540–553. https://doi.org/10.1038/nrmicro1662

Levitz SM, Huang H, Ostroff GR, Specht CA (2015) Exploiting fungal cell wall components in vaccines. Semin Immunopathol 37(2):199–207. https://doi.org/10.1007/s00281-014-0460-6

Liu X, Kwon YU, Seeberger PH (2005) Convergent synthesis of a fully lipidated glycosylphosphatidylinositol anchor of *Plasmodium falciparum*. J Am Chem Soc 127(14):5004–5005. https://doi.org/10.1021/ja042374o

Liu M, Capilla J, Johansen ME, Alvarado D, Martinez M, Chen V, Clemons KV, Stevens DA (2011) Saccharomyces as a vaccine against systemic aspergillosis: 'the friend of man' a friend again? J Med Microbiol 60(Pt 10):1423–1432. https://doi.org/10.1099/jmm.0.033290-0

Liu MCK, Johansen ME, Martinez M, Chen V, Stevens DA (2012) *Saccharomyces* as a vaccine against systemic candidiasis. Immunol Invest 41(8):847–855

Luo G, Ibrahim AS, Spellberg B, Nobile CJ, Mitchell AP, Fu Y (2010) *Candida albicans* Hyr1p confers resistance to neutrophil killing and is a potential vaccine target. J Infect Dis 201 (11):1718–1728. https://doi.org/10.1086/652407

Luong M, Lam JS, Chen J, Levitz SM (2007) Effects of fungal N- and O-linked mannosylation on the immunogenicity of model vaccines. Vaccine 25(22):4340–4344. https://doi.org/10.1016/j.vaccine.2007.03.027

Luyai AE, Heimburg-Molinaro J, Prasanphanich NS, Mickum ML, Lasanajak Y, Song X, Nyame AK, Wilkins P, Rivera-Marrero CA, Smith DF, Van Die I, Secor WE, Cummings RD (2014) Differential expression of anti-glycan antibodies in schistosome-infected humans, rhesus monkeys and mice. Glycobiology 24(7):602–618. https://doi.org/10.1093/glycob/cwu029

Macher BA, Galili U (2008) The Galalpha 1,3Galbeta1,4GlcNAc-R (alpha-Gal) epitope: a carbohydrate of unique evolution and clinical relevance. Biochim Biophys Acta 1780(2):75–88. https://doi.org/10.1016/j.bbagen.2007.11.003

MacLennan IC (1994) Germinal centers. Annu Rev Immunol 12:117–139. https://doi.org/10.1146/annurev.iy.12.040194.001001

Madhi SA, Kuwanda L, Cutland C, Klugman KP (2005) The impact of a 9-valent pneumococcal conjugate vaccine on the public health burden of pneumonia in HIV-infected and -uninfected children. Clin Infect Dis 40(10):1511–1518. https://doi.org/10.1086/429828

Majumder T, Liu M, Chen V, Martinez M, Alvarado D, Clemons KV, Stevens DA (2014) Killed *Saccharomyces cerevisiae* protects against lethal challenge of *Cryptococcus grubii*. Mycopathologia 178(3–4):189–195. https://doi.org/10.1007/s11046-014-9798-5

Mandalasi M, Dorabawila N, Smith DF, Heimburg-Molinaro J, Cummings RD, Nyame AK (2013) Development and characterization of a specific IgG monoclonal antibody toward the Lewis x antigen using splenocytes of Schistosoma mansoni-infected mice. Glycobiology 23(7):877–892. https://doi.org/10.1093/glycob/cwt025

McLellan JS, Pancera M, Carrico C, Gorman J, Julien JP, Khayat R, Louder R, Pejchal R, Sastry M, Dai K, O'Dell S, Patel N, Shahzad-ul-Hussan S, Yang Y, Zhang B, Zhou T, Zhu J, Boyington JC, Chuang GY, Diwanji D, Georgiev I, Kwon YD, Lee D, Louder MK, Moquin S, Schmidt SD, Yang ZY, Bonsignori M, Crump JA, Kapiga SH, Sam NE, Haynes BF, Burton DR, Koff WC, Walker LM, Phogat S, Wyatt R, Orwenyo J, Wang LX, Arthos J, Bewley CA, Mascola JR, Nabel GJ, Schief WR, Ward AB, Wilson IA, Kwong PD (2011) Structure of HIV-1 gp120 V1/V2 domain with broadly neutralizing antibody PG9. Nature 480 (7377):336–343. https://doi.org/10.1038/nature10696

Minor PD (2015) Live attenuated vaccines: historical successes and current challenges. Virology 479–480:379–392. https://doi.org/10.1016/j.virol.2015.03.032

Moldt B, Rakasz EG, Schultz N, Chan-Hui PY, Swiderek K, Weisgrau KL, Piaskowski SM, Bergman Z, Watkins DI, Poignard P, Burton DR (2012) Highly potent HIV-specific antibody neutralization in vitro translates into effective protection against mucosal SHIV challenge in vivo. Proc Natl Acad Sci U S A 109(46):18921–18925. https://doi.org/10.1073/pnas. 1214785109

Moser C, Muller M, Kaeser MD, Weydemann U, Amacker M (2013) Influenza virosomes as vaccine adjuvant and carrier system. Expert Rev Vaccines 12(7):779–791. https://doi.org/10. 1586/14760584.2013.811195

Mouquet H (2015) Tailored immunogens for rationally designed antibody-based HIV-1 vaccines. Trends Immunol 36(8):437–439. https://doi.org/10.1016/j.it.2015.07.001

Mouquet H, Scharf L, Euler Z, Liu Y, Eden C, Scheid JF, Halper-Stromberg A, Gnanapragasam PN, Spencer DI, Seaman MS, Schuitemaker H, Feizi T, Nussenzweig MC, Bjorkman PJ (2012) Complex-type N-glycan recognition by potent broadly neutralizing HIV antibodies. Proc Natl Acad Sci U S A 109(47):E3268–E3277. https://doi.org/10.1073/pnas. 1217207109

Moxon ER, Kroll JS (1990) The role of bacterial polysaccharides capsules as virulence factors. In: Jann K, Jann B (eds) Bacterial capsules. pp 65–85

Mukherjee J, Casadevall A, Scharff MD (1993) Molecular characterization of the humoral responses to *Cryptococcus neoformans* infection and glucuronoxylomannan-tetanus toxoid conjugate immunization. J Exp Med 177(4):1105–1116

Mukherjee J, Nussbaum G, Scharff MD, Casadevall A (1995) Protective and nonprotective monoclonal antibodies to *Cryptococcus neoformans* originating from one B cell. J Exp Med 181(1):405–409

Nordoy T, Aaberge IS, Husebekk A, Samdal HH, Steinert S, Melby H, Kolstad A (2002) Cancer patients undergoing chemotherapy show adequate serological response to vaccinations against influenza virus and *Streptococcus pneumoniae*. Med Oncol 19(2):71–78. https://doi.org/10. 1385/MO:19:2:71

Noriega FR, Liao FM, Maneval DR, Ren S, Formal SB, Levine MM (1999) Strategy for cross-protection among Shigella flexneri serotypes. Infect Immun 67(2):782–788

Nyame AK, Kawar ZS, Cummings RD (2004) Antigenic glycans in parasitic infections: implications for vaccines and diagnostics. Arch Biochem Biophys 426(2):182–200. https://doi. org/10.1016/j.abb.2004.04.004

Okano M, Satoskar AR, Nishizaki K, Abe M, Harn DA Jr (1999) Induction of Th2 responses and IgE is largely due to carbohydrates functioning as adjuvants on Schistosoma mansoni egg antigens. J Immunol 163(12):6712–6717

Pancera M, Shahzad-Ul-Hussan S, Doria-Rose NA, McLellan JS, Bailer RT, Dai K, Loesgen S, Louder MK, Staupe RP, Yang Y, Zhang B, Parks R, Eudailey J, Lloyd KE, Blinn J, Alam SM, Haynes BF, Amin MN, Wang LX, Burton DR, Koff WC, Nabel GJ, Mascola JR, Bewley CA, Kwong PD (2013) Structural basis for diverse N-glycan recognition by HIV-1-neutralizing

V1-V2-directed antibody PG16. Nat Struct Mol Biol 20(7):804–813. https://doi.org/10.1038/nsmb.2600

Pappas PG (2006) Invasive candidiasis. Infect Dis Clin North Am 20(3):485–506. https://doi.org/10.1016/j.idc.2006.07.004

Passwell JH, Ashkenzi S, Banet-Levi Y, Ramon-Saraf R, Farzam N, Lerner-Geva L, Even-Nir H, Yerushalmi B, Chu C, Shiloach J, Robbins JB, Schneerson R, Israeli Shigella Study G (2010) Age-related efficacy of Shigella O-specific polysaccharide conjugates in 1–4-year-old Israeli children. Vaccine 28(10):2231–2235. https://doi.org/10.1016/j.vaccine.2009.12.050

Phalipon A, Tanguy M, Grandjean C, Guerreiro C, Belot F, Cohen D, Sansonetti PJ, Mulard LA (2009) A synthetic carbohydrate-protein conjugate vaccine candidate against *Shigella flexneri* 2a infection. J Immunol 182(4):2241–2247. https://doi.org/10.4049/jimmunol.0803141

Pietrella D, Rachini A, Torosantucci A, Chiani P, Brown AJ, Bistoni F, Costantino P, Mosci P, d'Enfert C, Rappuoli R, Cassone A, Vecchiarelli A (2010) A beta-glucan-conjugate vaccine and anti-beta-glucan antibodies are effective against murine vaginal candidiasis as assessed by a novel in vivo imaging technique. Vaccine 28(7):1717–1725. https://doi.org/10.1016/j.vaccine.2009.12.021

Pitisuttithum P, Gilbert P, Gurwith M, Heyward W, Martin M, van Griensven F, Hu D, Tappero JW, Choopanya K, Bangkok Vaccine Evaluation G (2006) Randomized, double-blind, placebo-controlled efficacy trial of a bivalent recombinant glycoprotein 120 HIV-1 vaccine among injection drug users in Bangkok. Thailand. J Infect Dis 194(12):1661–1671. https://doi.org/10.1086/508748

Plaine A, Walker L, Da Costa G, Mora-Montes HM, McKinnon A, Gow NA, Gaillardin C, Munro CA, Richard ML (2008) Functional analysis of *Candida albicans* GPI-anchored proteins: roles in cell wall integrity and caspofungin sensitivity. Fungal Genet Biol 45(10):1404–1414. https://doi.org/10.1016/j.fgb.2008.08.003

Pozsgay V (2008) Recent developments in synthetic oligosaccharide-based bacterial vaccines. Curr Top Med Chem 8(2):126–140

Pozsgay V, Chu C, Pannell L, Wolfe J, Robbins JB, Schneerson R (1999) Protein conjugates of synthetic saccharides elicit higher levels of serum IgG lipopolysaccharide antibodies in mice than do those of the O-specific polysaccharide from Shigella dysenteriae type 1. Proc Natl Acad Sci U S A 96(9):5194–5197

Prasanphanich NS, Luyai AE, Song X, Heimburg-Molinaro J, Mandalasi M, Mickum M, Smith DF, Nyame AK, Cummings RD (2014) Immunization with recombinantly expressed glycan antigens from *Schistosoma mansoni* induces glycan-specific antibodies against the parasite. Glycobiology 24(7):619–637. https://doi.org/10.1093/glycob/cwu027

Prior JL, Prior RG, Hitchen PG, Diaper H, Griffin KF, Morris HR, Dell A, Titball RW (2003) Characterization of the O antigen gene cluster and structural analysis of the O antigen of *Francisella tularensis* subsp. *tularensis*. J Med Microbiol 52(Pt 10):845–851. https://doi.org/10.1099/jmm.0.05184-0

Pritchard LK, Vasiljevic S, Ozorowski G, Seabright GE, Cupo A, Ringe R, Kim HJ, Sanders RW, Doores KJ, Burton DR, Wilson IA, Ward AB, Moore JP, Crispin M (2015) Structural constraints determine the glycosylation of HIV-1 envelope trimers. Cell Rep 11(10):1604–1613. https://doi.org/10.1016/j.celrep.2015.05.017

Radosevic K, Rodriguez A, Mintardjo R, Tax D, Bengtsson KL, Thompson C, Zambon M, Weverling GJ, Uytdehaag F, Goudsmit J (2008) Antibody and T-cell responses to a virosomal adjuvanted H9N2 avian influenza vaccine: impact of distinct additional adjuvants. Vaccine 26(29–30):3640–3646. https://doi.org/10.1016/j.vaccine.2008.04.071

Richter J, Correia Dacal AR, Vergetti Siqueira JG, Poggensee G, Mannsmann U, Deelder A, Feldmeier H (1998) Sonographic prediction of variceal bleeding in patients with liver fibrosis due to *Schistosoma mansoni*. Trop Med Int Health 3(9):728–735

Riedel S (2005) Edward Jenner and the history of smallpox and vaccination. Proc (Bayl Univ Med Cent) 18(1):21–25

Rodrigues JA, Acosta-Serrano A, Aebi M, Ferguson MA, Routier FH, Schiller I, Soares S, Spencer D, Titz A, Wilson IB, Izquierdo L (2015) Parasite glycobiology: a Bittersweet Symphony. PLoS Pathog 11(11):e1005169. https://doi.org/10.1371/journal.ppat.1005169

Romani L (2011) Immunity to fungal infections. Nat Rev Immunol 11(4):275–288. https://doi.org/10.1038/nri2939

Rouvinski A, Guardado-Calvo P, Barba-Spaeth G, Duquerroy S, Vaney MC, Kikuti CM, Navarro Sanchez ME, Dejnirattisai W, Wongwiwat W, Haouz A, Girard-Blanc C, Petres S, Shepard WE, Despres P, Arenzana-Seisdedos F, Dussart P, Mongkolsapaya J, Screaton GR, Rey FA (2015) Recognition determinants of broadly neutralizing human antibodies against dengue viruses. Nature 520(7545):109–113. https://doi.org/10.1038/nature14130

Roy RM, Klein BS (2012) Dendritic cells in antifungal immunity and vaccine design. Cell Host Microbe 11(5):436–446. https://doi.org/10.1016/j.chom.2012.04.005

Royet J, Dziarski R (2007) Peptidoglycan recognition proteins: pleiotropic sensors and effectors of antimicrobial defences. Nat Rev Microbiol 5(4):264–277. https://doi.org/10.1038/nrmicro1620

Rubens CE, Wessels MR, Heggen LM, Kasper DL (1987) Transposon mutagenesis of type III group B Streptococcus: correlation of capsule expression with virulence. Proc Natl Acad Sci U S A 84(20):7208–7212

Rudd PM, Dwek RA (1997) Glycosylation: heterogeneity and the 3D structure of proteins. Crit Rev Biochem Mol Biol 32(1):1–100. https://doi.org/10.3109/10409239709085144

Sadanand S, Suscovich TJ, Alter G (2016) Broadly neutralizing antibodies against HIV: new insights to inform vaccine design. Annu Rev Med 67:185–200. https://doi.org/10.1146/annurev-med-091014-090749

Sanders RW, Venturi M, Schiffner L, Kalyanaraman R, Katinger H, Lloyd KO, Kwong PD, Moore JP (2002a) The mannose-dependent epitope for neutralizing antibody 2G12 on human immunodeficiency virus type 1 glycoprotein gp120. J Virol 76(14):7293–7305

Sanders RW, Vesanen M, Schuelke N, Master A, Schiffner L, Kalyanaraman R, Paluch M, Berkhout B, Maddon PJ, Olson WC, Lu M, Moore JP (2002b) Stabilization of the soluble, cleaved, trimeric form of the envelope glycoprotein complex of human immunodeficiency virus type 1. J Virol 76(17):8875–8889

Sanders RW, Derking R, Cupo A, Julien JP, Yasmeen A, de Val N, Kim HJ, Blattner C, de la Pena AT, Korzun J, Golabek M, de Los Reyes K, Ketas TJ, van Gils MJ, King CR, Wilson IA, Ward AB, Klasse PJ, Moore JP (2013) A next-generation cleaved, soluble HIV-1 Env trimer, BG505 SOSIP.664 gp140, expresses multiple epitopes for broadly neutralizing but not non-neutralizing antibodies. PLoS Pathog 9(9):e1003618. https://doi.org/10.1371/journal.ppat.1003618

Sanders RW, van Gils MJ, Derking R, Sok D, Ketas TJ, Burger JA, Ozorowski G, Cupo A, Simonich C, Goo L, Arendt H, Kim HJ, Lee JH, Pugach P, Williams M, Debnath G, Moldt B, van Breemen MJ, Isik G, Medina-Ramirez M, Back JW, Koff WC, Julien JP, Rakasz EG, Seaman MS, Guttman M, Lee KK, Klasse PJ, LaBranche C, Schief WR, Wilson IA, Overbaugh J, Burton DR, Ward AB, Montefiori DC, Dean H, Moore JP (2015) HIV-1 Vaccines. HIV-1 neutralizing antibodies induced by native-like envelope trimers. Science 349 (6244):aac4223. https://doi.org/10.1126/science.aac4223

Saville SP, Lazzell AL, Chaturvedi AK, Monteagudo C, Lopez-Ribot JL (2009) Efficacy of a genetically engineered *Candida albicans* tet-NRG1 strain as an experimental live attenuated vaccine against hematogenously disseminated candidiasis. Clin Vaccine Immunol 16(3):430–432. https://doi.org/10.1128/CVI.00480-08

Scanlan CN, Pantophlet R, Wormald MR, Ollmann Saphire E, Stanfield R, Wilson IA, Katinger H, Dwek RA, Rudd PM, Burton DR (2002) The broadly neutralizing anti-human immunodeficiency virus type 1 antibody 2G12 recognizes a cluster of alpha1→ 2 mannose residues on the outer face of gp120. J Virol 76(14):7306–7321

Scanlan CN, Offer J, Zitzmann N, Dwek RA (2007) Exploiting the defensive sugars of HIV-1 for drug and vaccine design. Nature 446(7139):1038–1045. https://doi.org/10.1038/nature05818

Scharf L, Wang H, Gao H, Chen S, McDowall AW, Bjorkman PJ (2015) Broadly neutralizing Antibody 8ANC195 recognizes closed and open states of HIV-1 Env. Cell 162(6):1379–1390. https://doi.org/10.1016/j.cell.2015.08.035

Scheid JF, Mouquet H, Feldhahn N, Seaman MS, Velinzon K, Pietzsch J, Ott RG, Anthony RM, Zebroski H, Hurley A, Phogat A, Chakrabarti B, Li Y, Connors M, Pereyra F, Walker BD, Wardemann H, Ho D, Wyatt RT, Mascola JR, Ravetch JV, Nussenzweig MC (2009) Broad diversity of neutralizing antibodies isolated from memory B cells in HIV-infected individuals. Nature 458(7238):636–640. https://doi.org/10.1038/nature07930

Schmidt CS, White CJ, Ibrahim AS, Filler SG, Fu Y, Yeaman MR, Edwards JE Jr, Hennessey JP Jr (2012) NDV-3, a recombinant alum-adjuvanted vaccine for *Candida* and *Staphylococcus aureus*, is safe and immunogenic in healthy adults. Vaccine 30(52):7594–7600. https://doi.org/10.1016/j.vaccine.2012.10.038

Schofield L, Novakovic S, Gerold P, Schwarz RT, McConville MJ, Tachado SD (1996) Glycosylphosphatidylinositol toxin of Plasmodium up-regulates intercellular adhesion molecule-1, vascular cell adhesion molecule-1, and E-selectin expression in vascular endothelial cells and increases leukocyte and parasite cytoadherence via tyrosine kinase-dependent signal transduction. J Immunol 156(5):1886–1896

Schofield L, Hewitt MC, Evans K, Siomos MA, Seeberger PH (2002) Synthetic GPI as a candidate anti-toxic vaccine in a model of malaria. Nature 418(6899):785–789. https://doi.org/10.1038/nature00937

Seaman MS, Janes H, Hawkins N, Grandpre LE, Devoy C, Giri A, Coffey RT, Harris L, Wood B, Daniels MG, Bhattacharya T, Lapedes A, Polonis VR, McCutchan FE, Gilbert PB, Self SG, Korber BT, Montefiori DC, Mascola JR (2010) Tiered categorization of a diverse panel of HIV-1 Env pseudoviruses for assessment of neutralizing antibodies. J Virol 84(3):1439–1452. https://doi.org/10.1128/JVI.02108-09

Shingai M, Donau OK, Plishka RJ, Buckler-White A, Mascola JR, Nabel GJ, Nason MC, Montefiori D, Moldt B, Poignard P, Diskin R, Bjorkman PJ, Eckhaus MA, Klein F, Mouquet H, Cetrulo Lorenzi JC, Gazumyan A, Burton DR, Nussenzweig MC, Martin MA, Nishimura Y (2014) Passive transfer of modest titers of potent and broadly neutralizing anti-HIV monoclonal antibodies block SHIV infection in macaques. J Exp Med 211(10):2061–2074. https://doi.org/10.1084/jem.20132494

Shoham S, Huang C, Chen JM, Golenbock DT, Levitz SM (2001) Toll-like receptor 4 mediates intracellular signaling without TNF-alpha release in response to Cryptococcus neoformans polysaccharide capsule. J Immunol 166(7):4620–4626

Smit CH, Homann A, van Hensbergen VP, Schramm G, Haas H, van Diepen A, Hokke CH (2015) Surface expression patterns of defined glycan antigens change during *Schistosoma mansoni* cercarial transformation and development of schistosomula. Glycobiology 25(12):1465–1479. https://doi.org/10.1093/glycob/cwv066

Sok D, Doores KJ, Briney B, Le KM, Saye-Francisco KL, Ramos A, Kulp DW, Julien JP, Menis S, Wickramasinghe L, Seaman MS, Schief WR, Wilson IA, Poignard P, Burton DR (2014) Promiscuous glycan site recognition by antibodies to the high-mannose patch of gp120 broadens neutralization of HIV. Sci Transl Med 6(236):236ra263. https://doi.org/10.1126/scitranslmed.3008104

Sommerstein R, Flatz L, Remy MM, Malinge P, Magistrelli G, Fischer N, Sahin M, Bergthaler A, Igonet S, Ter Meulen J, Rigo D, Meda P, Rabah N, Coutard B, Bowden TA, Lambert PH, Siegrist CA, Pinschewer DD (2015) Arenavirus glycan shield promotes neutralizing antibody evasion and protracted infection. PLoS Pathog 11(11):e1005276. https://doi.org/10.1371/journal.ppat.1005276

Specht CA, Nong S, Dan JM, Lee CK, Levitz SM (2007) Contribution of glycosylation to T cell responses stimulated by recombinant *Cryptococcus neoformans* mannoprotein. J Infect Dis 196 (5):796–800. https://doi.org/10.1086/520536

Spellberg B (2011) Vaccines for invasive fungal infections. F1000 Med Rep 3:13. https://doi.org/10.3410/m3-13

Spellberg B, Ibrahim AS, Lin L, Avanesian V, Fu Y, Lipke P, Otoo H, Ho T, Edwards JE Jr (2008) Antibody titer threshold predicts anti-candidal vaccine efficacy even though the mechanism of protection is induction of cell-mediated immunity. J Infect Dis 197(7):967–971. https://doi.org/10.1086/529204

Srinivasan S, Ghosh M, Maity S, Varadarajan R (2016) Broadly neutralizing antibodies for therapy of viral infections. Antibody Technol J 6:1–15

Steichen JM, Kulp DW, Tokatlian T, Escolano A, Dosenovic P, Stanfield RL, McCoy LE, Ozorowski G, Hu X, Kalyuzhniy O, Briney B, Schiffner T, Garces F, Freund NT, Gitlin AD, Menis S, Georgeson E, Kubitz M, Adachi Y, Jones M, Mutafyan AA, Yun DS, Mayer CT, Ward AB, Burton DR, Wilson IA, Irvine DJ, Nussenzweig MC, Schief WR (2016) HIV vaccine design to target germline precursors of glycan-dependent broadly neutralizing antibodies. Immunity 45(3):483–496. https://doi.org/10.1016/j.immuni.2016.08.016

Sun JB, Stadecker MJ, Mielcarek N, Lakew M, Li BL, Hernandez HJ, Czerkinsky C, Holmgren J (2001) Nasal administration of *Schistosoma mansoni* egg antigen-cholera B subunit conjugate suppresses hepatic granuloma formation and reduces mortality in *S. mansoni*-infected mice. Scand J Immunol 54(5):440–447

Tachado SD, Gerold P, McConville MJ, Baldwin T, Quilici D, Schwarz RT, Schofield L (1996) Glycosylphosphatidylinositol toxin of *Plasmodium* induces nitric oxide synthase expression in macrophages and vascular endothelial cells by a protein tyrosine kinase-dependent and protein kinase C-dependent signaling pathway. J Immunol 156(5):1897–1907

Tedaldi EM, Baker RK, Moorman AC, Wood KC, Fuhrer J, McCabe RE, Holmberg SD, Investigators HIVOS (2004) Hepatitis A and B vaccination practices for ambulatory patients infected with HIV. Clin Infect Dis 38(10):1478–1484. https://doi.org/10.1086/420740

Terra VS, Mills DC, Yates LE, Abouelhadid S, Cuccui J, Wren BW (2012) Recent developments in bacterial protein glycan coupling technology and glycoconjugate vaccine design. J Med Microbiol 61(Pt 7):919–926. https://doi.org/10.1099/jmm.0.039438-0

Thomas PG, Harn DA Jr (2004) Immune biasing by helminth glycans. Cell Microbiol 6(1):13–22

Todryk SM, Hill AV (2007) Malaria vaccines: the stage we are at. Nat Rev Microbiol 5(7):487–489. https://doi.org/10.1038/nrmicro1712

Torosantucci A, Bromuro C, Chiani P, De Bernardis F, Berti F, Galli C, Norelli F, Bellucci C, Polonelli L, Costantino P, Rappuoli R, Cassone A (2005) A novel glyco-conjugate vaccine against fungal pathogens. J Exp Med 202(5):597–606. https://doi.org/10.1084/jem.20050749

US Food and Drug Administration (2015) Complete list of vaccines licensed for immunization and distribution in the US. http://www.fda.gov/BiologicsBloodVaccines/Vaccines/ApprovedProducts/ucm093833.htm

van Dam JE, Fleer A, Snippe H (1990) Immunogenicity and immunochemistry of *Streptococcus pneumoniae* capsular polysaccharides. Antonie Van Leeuwenhoek 58(1):1–47

van der Put RM, Kim TH, Guerreiro C, Thouron F, Hoogerhout P, Sansonetti PJ, Westdijk J, Stork M, Phalipon A, Mulard LA (2016) A synthetic carbohydrate conjugate vaccine candidate against shigellosis: improved bioconjugation and impact of alum on immunogenicity. Bioconjug Chem 27(4):883–892. https://doi.org/10.1021/acs.bioconjchem.5b00617

van Diepen A, Smit CH, van Egmond L, Kabatereine NB, Pinot de Moira A, Dunne DW, Hokke CH (2012a) Differential anti-glycan antibody responses in Schistosoma mansoni-infected children and adults studied by shotgun glycan microarray. PLoS Negl Trop Dis 6(11):e1922. https://doi.org/10.1371/journal.pntd.0001922

van Diepen A, Van der Velden NS, Smit CH, Meevissen MH, Hokke CH (2012b) Parasite glycans and antibody-mediated immune responses in Schistosoma infection. Parasitology 139(9):1219–1230. https://doi.org/10.1017/S0031182012000273

van Diepen A, van der Plas AJ, Kozak RP, Royle L, Dunne DW, Hokke CH (2015) Development of a *Schistosoma mansoni* shotgun O-glycan microarray and application to the discovery of new antigenic schistosome glycan motifs. Int J Parasitol 45(7):465–475. https://doi.org/10.1016/j.ijpara.2015.02.008

Van Roon AM, Van de Vijver KK, Jacobs W, Van Marck EA, Van Dam GJ, Hokke CH, Deelder AM (2004) Discrimination between the anti-monomeric and the anti-multimeric Lewis X response in murine schistosomiasis. Microbes Infect 6(13):1125–1132. https://doi.org/10.1016/j.micinf.2004.06.004

Vecchiarelli A (2000) Immunoregulation by capsular components of *Cryptococcus neoformans*. Med Mycol 38(6):407–417

Vigerust DJ, Shepherd VL (2007) Virus glycosylation: role in virulence and immune interactions. Trends Microbiol 15(5):211–218. https://doi.org/10.1016/j.tim.2007.03.003

Wacker M, Feldman MF, Callewaert N, Kowarik M, Clarke BR, Pohl NL, Hernandez M, Vines ED, Valvano MA, Whitfield C, Aebi M (2006) Substrate specificity of bacterial oligosaccharyltransferase suggests a common transfer mechanism for the bacterial and eukaryotic systems. Proc Natl Acad Sci U S A 103(18):7088–7093. https://doi.org/10.1073/pnas.0509207103

Walker LM, Huber M, Doores KJ, Falkowska E, Pejchal R, Julien JP, Wang SK, Ramos A, Chan-Hui PY, Moyle M, Mitcham JL, Hammond PW, Olsen OA, Phung P, Fling S, Wong CH, Phogat S, Wrin T, Simek MD, Protocol GPI, Koff WC, Wilson IA, Burton DR, Poignard P (2011) Broad neutralization coverage of HIV by multiple highly potent antibodies. Nature 477 (7365):466–470. https://doi.org/10.1038/nature10373

Wang LX (2013) Synthetic carbohydrate antigens for HIV vaccine design. Curr Opin Chem Biol 17(6):997–1005

Warburg A, Shtern A, Cohen N, Dahan N (2007) Laminin and a Plasmodium ookinete surface protein inhibit melanotic encapsulation of Sephadex beads in the hemocoel of mosquitoes. Microbes Infect 9(2):192–199. https://doi.org/10.1016/j.micinf.2006.11.006

WHO (2015) Immunization, vaccines and biologicals—WHO recommendations for routine immunization—summary tables

Wyle FA, Artenstein MS, Brandt BL, Tramont EC, Kasper DL, Altieri PL, Berman SL, Lowenthal JP (1972) Immunologic response of man to group B meningococcal polysaccharide vaccines. J Infect Dis 126(5):514–521

Xin H, Dziadek S, Bundle DR, Cutler JE (2008) Synthetic glycopeptide vaccines combining beta-mannan and peptide epitopes induce protection against candidiasis. Proc Natl Acad Sci U S A 105(36):13526–13531. https://doi.org/10.1073/pnas.0803195105

Yang Z, Li J, Liu Q, Yuan T, Zhang Y, Chen LQ, Lou Q, Sun Z, Ying H, Xu J, Dimitrov DS, Zhang MY (2015) Identification of non-HIV immunogens that bind to Germline b12 predecessors and prime for elicitation of cross-clade neutralizing HIV-1 antibodies. PLoS ONE 10(5):e0126428. https://doi.org/10.1371/journal.pone.0126428

Yilmaz B, Portugal S, Tran TM, Gozzelino R, Ramos S, Gomes J, Regalado A, Cowan PJ, d'Apice AJ, Chong AS, Doumbo OK, Traore B, Crompton PD, Silveira H, Soares MP (2014) Gut microbiota elicits a protective immune response against malaria transmission. Cell 159 (6):1277–1289. https://doi.org/10.1016/j.cell.2014.10.053

Zhao L, Seth A, Wibowo N, Zhao CX, Mitter N, Yu C, Middelberg AP (2014) Nanoparticle vaccines. Vaccine 32(3):327–337. https://doi.org/10.1016/j.vaccine.2013.11.069

Zhou J, Lottenbach KR, Barenkamp SJ, Lucas AH, Reason DC (2002) Recurrent variable region gene usage and somatic mutation in the human antibody response to the capsular polysaccharide of *Streptococcus pneumoniae* type 23F. Infect Immun 70(8):4083–4091

B Cells Carrying Antigen Receptors Against Microbes as Tools for Vaccine Discovery and Design

Deepika Bhullar and David Nemazee

Contents

Abstract Can basic science improve the art of vaccinology? Here, we review efforts to understand immune responses with the aim to improve vaccine design and, eventually, to predict the efficacy of human vaccine candidates using the tools of transformed B cells and targeted transgenic mice carrying B cells with antigen receptors specific for microbes of interest.

1 Introduction

Vaccinology is, and has always been, a crucial field for public health and for basic research. Early vaccine studies led to the discovery of antibodies, which were identified by their ability to neutralize microbial toxins upon passive transfer to naive animals. While some well-known vaccines are capable of eradicating important

D. Bhullar · D. Nemazee (✉)
Department of Immunology and Microbiology, The Scripps Research Institute,
10550 North Torrey Pines Rd, IM29, La Jolla, CA 92037, USA
e-mail: nemazee@scripps.edu

Current Topics in Microbiology and Immunology (2020) 428: 165–180
https://doi.org/10.1007/82_2019_156
© Springer Nature Switzerland AG 2019
Published Online: 28 March 2019

human pathogens, we lack adequate vaccines for many, perhaps most, human pathogens. For these challenging pathogens, the classical empirical approach to vaccinology fails, necessitating better knowledge of the microbes and the basic science of successful immune responses.

When a vaccine does not work, it is often difficult to determine the reason. Basic studies in B cell biology can have an impact on finding a way forward. One facet of this approach is in the discovery of good vaccine targets. The isolation of neutralizing antibodies derived from the B cells of infected patients through the use of methodologies such as phage display, hybridoma technology, and single-cell antibody gene cloning has identified crucial epitopes toward which one can focus vaccine responses. As described in the accompanying chapter by Dennis Burton and colleagues, this approach has been fruitful in the identification of neutralizing epitopes to pathogens with high diversity such as HIV, which require broadly neutralizing antibodies (bnAbs) that neutralize many substrains.

2 Recurrent Clonotypes

Besides their value in identifying epitopes of vulnerability, bnAb sequences provide information about the reproducibility of desired antibody responses. In many cases where antibody responses to defined epitopes have been studied in detail, certain VH or VL genes and often VH/VL pairs are recurrently selected (Nisonoff and Ju 1976; Crews et al. 1981; Wysocki et al. 1985; Kaartinen et al. 1983; Gearhart et al. 1981). This is most apparent in inbred animals, where the appearance of recurrent "clonotypes" was apparent many years ago from isoelectric focusing and idiotype studies (Sigal et al. 1977; Sher and Cohn 1972; Lieberman et al. 1974). In early studies in mice, certain responses were useful in mapping VH gene alleles that conferred specificity, for example, to microbial cell wall components such as phosphorylcholine (Lieberman et al. 1974). A particular VL gene was also associated with the anti-influenza A response of mice (Clarke et al. 1990). It was appreciated that recurrent responses were associated with high precursor frequency (Sigal et al. 1977) and sometimes microbial resistance (Mi et al. 2000). In the lightly mutated human antibody responses to bacteria, recurrent responses have been identified for the capsular polysaccharides of *Streptococcus pneumoniae* 23F (Zhou et al. 2002) and *Haemophilus influenzae* (Lucas and Reason 1999). In the case of Hib, lack of a particular VK gene has been linked to disease susceptibility (Feeney et al. 1996; Nadel et al. 1998). Human neutralizing antibodies to the cytomegalovirus AD-2S1 epitope appear to use lightly mutated versions of a single VH/VL pair and make contacts largely with non-mutated residues (Thomson et al. 2008). Remarkably, the recurrent use of particular VH or VL genes also occurs in certain classes of heavily mutated human bnAbs to highly diverse microbes, including the CD4 binding site of HIV (Zhou et al. 2013; Wu et al. 2011;

Bonsignori et al. 2012; Zhou et al. 2010) and the stem region of influenza hemagglutinin (Sui et al. 2009; Ekiert et al. 2011; Wrammert et al. 2011; Whittle et al. 2011; Dreyfus et al. 2012; Lingwood et al. 2012; Kashyap et al. 2008; Throsby et al. 2008; Whittle et al. 2014). Identification of V-gene recurrence likely requires analysis of well-defined epitopes. A recent deep sequencing study of dengue-exposed donors failed to find VH or VL dominance in the population as a whole but could identify CDRH3 motifs shared between independent infected donors, suggesting a convergent evolution in this case (Parameswaran et al. 2013).

2.1 From BnAb Sequences to "Germline" BnAb B Cells

The identification of reproducible bnAbs from humans also defines the receptor of the B cell making this desirable response. Although the typical bnAb sequence is mutated by activation-induced cytidine deaminase (AID)-catalyzed diversification, it is usually possible to infer the antibody sequence prior to mutation using sequence analysis programs (Gaeta et al. 2007; Alamyar et al. 2012; Ye et al. 2013; Russ et al. 2015). This so-called germline (gl)-bnAb in turn defines the B cell receptor (BCR) carried by the naive B cell giving rise to that bnAb. In a real sense, vaccines must target gl-bnAb BCRs and further promote and select their appropriate mutants, the bnAbs. Although these bnAbs can differ widely, as discussed above, reproducible responses are of particular interest for responses that are difficult to elicit. Accordingly, in vitro models of B cells carrying bnAbs and gl-bnAbs as BCRs have proven to be useful tools in vaccine research (Lingwood et al. 2012; Ota et al. 2012; Jardine et al. 2013, 2015; Hoot et al. 2013; McGuire et al. 2014a, b). Unlike free antibody, BCRs on the B cell surface are topologically constrained by the plasma membrane and associated with other cell surface molecules. An additional constraint in naive B cells is that IgM lacks the hinge region present in IgG antibodies. Activation by antigen of B cells carrying bnAb or gl-bnAb BCRs provides a stringent test for the ability of vaccine candidates or other antigens to stimulate B cells. B cells carrying bnAbs and gl-bnAbs as BCRs thus provide in vitro models to evaluate and design vaccine biologically active immunogens.

One way such models are generated experimentally is by the transfection of B cell tumor lines with the desired antibody genes carrying the membrane form of the H-chain. These cells have many similarities to the B cells found in vivo in their biochemical triggering through the BCR. Owing to polar residues in its transmembrane domain, the membrane form of antibody does not normally come to the cell surface unless associated with the signal transducer complex Ig-α/β (CD79a/CD79b) (Venkitaraman et al. 1991). In fact, B cell transfection experiments were critical in the discovery of Ig-α/β (Hombach et al. 1990). Ig-α and Ig-β are B cell-restricted transmembrane proteins carrying in their cytoplasmic domains the so-called ITAM (immunoreceptor tyrosine activation motif: $YxxL/Ix_{(6-8)}YxxL/I$)

common to many activating receptors in leukocyte biology, including the CD3 components of the T cell receptor (Reth 1989). These tyrosines become phosphorylated by src family kinases upon activation, leading to a cascade of events including recruitment of the tyrosine kinase Syk and the activation of additional enzymes and second messengers (Reth 1992). PLCγ in particular is responsible for initiating Ca^{++} mobilization, which is a convenient early readout. Later steps in activation in primary B cells include the upregulation of surface markers such as CD69 and CD86 (Cambier and Monroe 1984; Hara et al. 1986; Lenschow et al. 1994), which promote T cell interactions and whose upregulation is often also mimicked in transduced B cell lines stimulated by ligands that ligate the BCR.

3   B Cell Tumor Models for the Candidate Vaccine Antigen Response

A number of laboratories have used B cell lines transfected with vectors encoding HIV bnAbs BCRs to evaluate candidate vaccine antigens for bioactivity and to assess novel ligands for HIV gl-bnAb BCRs (Ota et al. 2012; Jardine et al. 2013; Jardine et al. 2015; Hoot et al. 2013; McGuire et al. 2014, 2013; Doores et al. 2013). These studies reinforced the finding for soluble IgG bnAbs that reversion of these mutated antibodies to the inferred germline sequence eliminates binding by demonstrating the lack of effective bioactivity. One surprise in these studies was that HIV virions were poorly stimulatory even to B cells carrying bnAb receptors (Ota et al. 2012) a result that was subsequently supported by studies in b12 transgenic mice (Ota et al. 2013). Less surprising is the fact that highly multimeric forms of antigen such as nanoparticles carrying repeating subunits, conjugates on virus-like particles, or liposome mounted antigens were most effective in vitro (Ota et al. 2012; Doores et al. 2013) (Ingale et al. in press). McGuire, Stamatatos, and colleagues have carried out extensive studies using gl-bnAb-expressing B cell lines to investigate the role of carbohydrate associated with HIV Env in limiting the functional access to bnAb and gl-bnAb BCRs on the B cell surface (Hoot et al. 2013; McGuire et al. 2014, 2013). An important conclusion from these studies was that certain wild-type envelopes could be recognized by gl-bnAb BCRs provided that one or more key N-glycosylation sites flanking the site of vulnerability were eliminated. Studies on cells carrying membrane-bound anti-influenza hemagglutinin antibodies have also been carried out (Lingwood et al. 2012; Weaver et al. 2016). Interestingly, these investigators were able to express membrane IgG bnAbs and gl-bnAbs on 293 cells, a non-lymphoid cell line that is easy to transfect and lacks Ig-α/β. It is unclear why the BCR is able to come to the plasma membrane in this context, and the cells cannot signal as in a B cell activation assay. Nonetheless, the system has been useful in assessing some aspects of BCR/antigen interactions (Lingwood et al. 2012; Weaver et al. 2016).

4 Immunoglobulin Transgenic and Knock-in Mice for Vaccine Research

Rearranged immunoglobulin genes were among the first genes to be used in the generation and study of transgenic mice (Brinster et al. 1983; Storb et al. 1986; Storb 1987). Technically, this is carried out using microinjection into the male pronucleus of a recently fertilized egg (zygote). These first-generation transgenics inserted the microinjected DNA randomly into the genome, usually in multicopy arrays, which led to varied and often nonphysiological expression patterns. Researchers quickly realized that the technology requires careful transgene design to include appropriate regulatory elements in cis and careful selection of transgenic lines with appropriate expression. The early studies, along with related knockout studies, supported a model of feedback suppression of antibody gene expression: expression in developing B cells of an active, pre-rearranged transgenic immunoglobulin gene would tend to suppress or prevent endogenous antibody gene rearrangement (Ritchie et al. 1984; Weaver et al. 1985; Nussenzweig et al. 1987; Kitamura and Rajewsky 1992; Rusconi and Kohler 1985; Hagman et al. 1989; Betz et al. 1993). Transgenes expressing antibody H/L pairs not only could lead to expression of predefined antibody to an antigen of interest, but also suppressed other specificities by promoting B cell development and blocking endogenous rearrangements (Rusconi and Kohler 1985), leading in some instances to mice with virtually monoclonal B cell populations (Goodnow et al. 1988; Nemazee and Burki 1989; Russell et al. 1991). These "conventional" antibody transgenic mice have proven to be very useful in the study of B cell development and self-tolerance (Goodnow et al. 1988; Nemazee and Burki 1989; Russell et al. 1991; Erikson et al. 1991; Arnold et al. 1994; Borrero and Clarke 2002; Carsetti et al. 1995; Kenny et al. 1991; Brink et al. 1992; Gay et al. 1993; Tiegs et al. 1993; Fulcher and Basten 1994; Hayakawa et al. 1999; Chumley et al. 2000; Hayakawa et al. 2003; Foster et al. 1997; Shlomchik et al. 1993) and in the response to microbial antigens such as LCMV, VSV, and influenza (Seiler et al. 1998; Martin et al. 2001; Carmack et al. 1991, 1990). However, these models had limitations for the analysis of immunity, such as an inability to undergo H-chain class switching (owing to a lack of downstream H-chain genes), and their multicopy nature, which made analysis of aspects such as somatic mutation difficult (Betz et al. 1993; O'Brien et al. 1987). The usefulness of these conventional Ig transgenic models for vaccine design has been mainly in aiding research into the T cell-independent immune response, in facilitating visualization of the responding cells, and in the analysis of bystander activation (Seiler et al. 1998; Senn et al. 2003).

4.1 Knock-in Mice

More recently, antibody gene "knock-in" mice was developed, in which the antibody transgenes of interest are targeted to the physiological locus (Chen et al. 1995; Taki et al. 1993; Luning Prak and Weigert 1995; Pelanda et al. 1997; Cascalho et al. 1996; Sonoda et al. 1997; Pewzner-Jung et al. 1998; Litzenburger et al. 2000; Phan et al. 2003; Hangartner et al. 2003; Berland et al. 2006; Hangartner et al. 2006). Targeting to the immunoglobulin locus provides more physiological genetic control, allowing such key features such as robust somatic mutation and class switching. However, despite this fairly good physiological control, the antibody genes introduced by targeting can be eliminated in developing B cells by physiological receptor editing and in preB cells by VH replacement (Chen et al. 1995; Luning Prak and Weigert 1995; Pelanda et al. 1996; Casellas et al. 2001) or by the nonphysiological use of the targeted VH element as an acceptor of DH invasion (Taki et al. 1993; Cascalho et al. 1996; Golub et al. 2001; Koralov et al. 2006). These latter phenomena involve the recombination by upstream VH or DH elements to a conserved heptamer signal sequence site within the knock-in coding region (TACTGTG), which is present in many germline VH regions of mouse and human. Such rearrangements are typically destructive, leading to expression of the alternate IgH allele (Chen et al. 1995; Luning Prak and Weigert 1995; Casellas et al. 2001; Taki et al. 1995). An upshot of these recombinations is that B cells in knock-in mice are rarely monoclonal, and in some extreme cases the transgene-encoded specificity is barely expressed (Pelanda et al. 1997; Chen et al. 1997). When the B cells are autoreactive, negative selection can occur by several mechanisms, including apoptosis, anergy, or receptor editing in the bone marrow (reviewed in) Nemazee 2006; Cambier et al. 2007; Shlomchik 2008; Goodnow et al. 2005. Receptor editing typically results in ongoing L-chain gene recombination, which can displace a functional L-chain gene or inactivate it, leading to its functional replacement.

Among the strengths of the knock-in technology is that it allows one to identify BCRs that fail to support B cell development. Developmental failure can occur if the BCR is sufficiently autoreactive or if the antibody chain in question has other structural defects that prevent proper folding or association with the partner chain. The effects of autoreactivity on B cell development have been extensively studied in models designed for the purpose (reviewed in) Nemazee 2006; however, increasing evidence suggests that some desirable, even broadly neutralizing, antibody specificities to HIV may be negatively selected (Verkoczy and Diaz 2014; Haynes et al. 2005; Verkoczy et al. 2011; Doyle-Cooper et al. 2013; Chen et al. 2013; Finton et al. 2013; Yang et al. 2013). Given the high safety standards for human vaccines, epitopes that require such negatively selected specificities might be undesirable. In the case of the well-known anti-HIV gp41 broadly neutralizing antibodies 2F5 and 4E10, it has been proposed that specific intracellular self-ligands promote negative selection (Finton et al. 2013; Yang et al. 2013). Definitive testing of this hypothesis should be possible by assessing the phenotype of 2F5 and 4E10 knock-in mice in which the cognate epitopes are eliminated, in which negative selection is predicted to be relieved.

Knock-in antibody mice have been useful in studies of microbial resistance, viral evasion, and vaccinology. Ig H-only or H/L knock-in models have been generated using antibodies specific for Streptococcus pneumoniae (Taki et al. 1995; Hu et al. 2002), VSV (Hangartner et al. 2003), LCMV (Hangartner et al. 2003; Hangartner et al. 2006), and HIV (Jardine et al. 2015; Ota et al. 2013; Verkoczy et al. 2011; Doyle-Cooper et al. 2013; Chen et al. 2013; Finton et al. 2013; Verkoczy et al. 2010; Verkoczy et al. 2013; Dosenovic et al. 2015; Zhang et al. 2016). Several of these studies involved the analysis of mice carrying antibody H-chains from neutralizing antibodies (usually mutated) paired with random mouse L-chains (Jardine et al. 2015; Ota et al. 2013; Hangartner et al. 2003; Hangartner et al. 2006; Finton et al. 2013; Hu et al. 2002; Dosenovic et al. 2015). Other studies involved knock-in mice expressing both H- and L-chains derived from neutralizing Abs (Ota et al. 2013; Verkoczy et al. 2011; Doyle-Cooper et al. 2013; Chen et al. 2013). More recently, mice have been generated to express H or H + L genes encoding the inferred non-mutated precursors of HIV broadly neutralizing antibodies (Jardine et al. 2015; Dosenovic et al. 2015; Zhang et al. 2016). These last allow one to assess the ability of experimental vaccination to mature the response appropriately, starting with a defined gene of known potential. These recent studies have indicated the outlines of a priming and possible booster vaccination pathway to elicit antibodies to the CD4 binding site on HIV Env (Jardine et al. 2015; Dosenovic et al. 2015). What makes these studies unique is that the priming immunogen used was targeted to a specific human gene VH1-2*02 and a particular length and sequence in the CDRL3 loop of L-chain. Testing such vaccines that are intended for human vaccination in small animals was not possible without a knock-in or comparable approach.

It is important to keep in mind some of the limitations of these knock-in models for vaccine research. A high precursor frequency of B cells expressing the BCR of a neutralizing antibody facilitates many analyses, but also provides B cells at super-physiological copy number. Moreover, in many models, some of the transgenic Ig is spontaneously secreted, which could affect certain analyses, such as by providing preimmune resistance to infection (Hangartner et al. 2006). Fortunately, these deficiencies can be readily overcome by transfer of limiting B cell numbers to adoptive recipients prior to vaccination (Ota et al. 2013; Hangartner et al. 2006).

4.2 Nuclear Transfer Mice

A more recent way to generate mouse models carrying defined receptors is to clone mice from the nuclei of lymphocytes (Hochedlinger and Jaenisch 2002; Kirak et al. 2010). This feat has been achieved many times by a small number of laboratories.

When the lymphocytes come from immunized individuals, the approach permits the isolation of antigen-specific cells (Kirak et al. 2010; Dougan et al. 2012). For example, Kirak et al. generated transnuclear mice with T cells specific for *Toxoplasma gondii*. The resulting mice have predefined receptors, though for B lymphocytes these are often "pre-switched" to downstream H-chain isotypes (Dougan et al. 2012; Kumar et al. 2015), and the receptors can still be modified by receptor editing (Gerdes and Wabl 2004) or VH replacement at the proB cell stage (Kumar et al. 2015). Although remarkably useful for a range of basic studies, the nuclear transfer approach is somewhat laborious and does not allow the type of precise pre-engineering of antibody sequence that is feasible in the knock-in approach.

5 Hu/SCID Mice in Vaccine Research

A final aspect of contemporary vaccine research concerns the increasing "humanization" of mouse models. One approach is to reconstitute immunodeficient mice with human hematopoietic cells. However, the humoral immune responses of such chimeras are so far suboptimal, which limits the use of these models to study the immune response to vaccination (Villaudy et al. 2014; Karpel et al. 2015). On the other hand, such models have proved to be useful in the analysis of passive antibody immunity and the mutational escape of microbes such as HIV (Karpel et al. 2015; Hur et al. 2012; Klein et al. 2012).

5.1 *Mice and Rats Carrying the Full Complement of Human Ig Genes*

An alternative approach that has proved fruitful in making humanized antibodies is the use of mice with inactivated endogenous antibody genes engineered to carry large transgenes encoding human immunoglobulin loci that are composed of many or all gene segments (reviewed in) Bruggemann et al. 2015. Immune responses in these animals seem to work most efficiently if the human H-chain VDJ elements are placed upstream of constant regions of the host (Pruzina et al. 2011; Osborn et al. 2013; Green 2014; Ma et al. 2013), presumably because the Fc portions of the antibodies interact properly with the mouse FcRs and complement. These models would be ideal for many human vaccine studies in which particular V genes are targeted for priming, as discussed above. As a practical matter, however, the mice in question are not easily accessible through normal scientific exchange, owing to their remarkable commercial value for the generation of monoclonal antibodies. And so these models have sadly had little impact so far in the basic science of vaccinology.

References

Alamyar E, Duroux P, Lefranc MP, Giudicelli V (2012) IMGT((R)) tools for the nucleotide analysis of immunoglobulin (IG) and T cell receptor (TR) V-(D)-J repertoires, polymorphisms, and IG mutations: IMGT/V-QUEST and IMGT/HighV-QUEST for NGS. Methods Mol Biol 882:569–604

Arnold LW, Pennell CA, McCray SK, Clarke SH (1994) Development of B-1 cells: segregation of phosphatidyl choline-specific B cells to the B-1 population occurs after immunoglobulin gene expression. J Exp Med 179:1585–1595

Berland R, Fernandez L, Kari E, Han JH, Lomakin I, Akira S, Wortis HH, Kearney JF, Ucci AA, Imanishi-Kari T (2006) Toll-like receptor 7-dependent loss of B cell tolerance in pathogenic autoantibody knockin mice. Immunity 25:429–440

Betz AG, Rada C, Pannell R, Milstein C, Neuberger MS (1993) Passenger transgenes reveal intrinsic specificity of the antibody hypermutation mechanism: clustering, polarity, and specific hot spots. Proc Natl Acad Sci USA 90:2385–2388

Bonsignori M, Montefiori DC, Wu X, Chen X, Hwang KK, Tsao CY, Kozink DM, Parks RJ, Tomaras GD, Crump JA, Kapiga SH, Sam NE, Kwong PD, Kepler TB, Liao HX, Mascola JR, Haynes BF (2012) Two distinct broadly neutralizing antibody specificities of different clonal lineages in a single HIV-1-infected donor: implications for vaccine design. J Virol 86:4688–4692

Borrero M, Clarke SH (2002) Low-affinity anti-Smith antigen B cells are regulated by anergy as opposed to developmental arrest or differentiation to B-1. J Immunol 168:13–21

Brink R, Goodnow CC, Crosbie J, Adams E, Eris J, Mason DY, Hartley SB, Basten A (1992) Immunoglobulin M and D antigen receptors are both capable of mediating B lymphocyte activation, deletion, or anergy after interaction with specific antigen. J Exp Med 176:991–1005

Brinster RL, Ritchie KA, Hammer RE, O'Brien RL, Arp B, Storb U (1983) Expression of a microinjected immunoglobulin gene in the spleen of transgenic mice. Nature 306:332–336

Bruggemann M, Osborn MJ, Ma B, Hayre J, Avis S, Lundstrom B, Buelow R (2015) Human antibody production in transgenic animals. Arch Immunol Ther Exp (Warsz) 63:101–108

Cambier JC, Monroe JG (1984) B cell activation. V. Differentiation signaling of B cell membrane depolarization, increased I-A expression, G0 to G1 transition, and thymidine uptake by anti-IgM and anti-IgD antibodies. J Immunol 133:576–581

Cambier JC, Gauld SB, Merrell KT, Vilen BJ (2007) B-cell anergy: from transgenic models to naturally occurring anergic B cells? Nat Rev Immunol 7:633–643

Carmack CE, Shinton SA, Hayakawa K, Hardy RR (1990) Rearrangement and selection of VH11 in the Ly-1 B cell lineage. J Exp Med 172:371–374

Carmack CE, Camper SA, Mackle JJ, Gerhard WU, Weigert Mg (1991) Influence of a V kappa 8 L chain transgene on endogenous rearrangements and the immune response to the HA(Sb) determinant on influenza virus. J Immunol 147, 2024–2033

Carsetti R, Kohler G, Lamers MC (1995) Transitional B cells are the target of negative selection in the B cell compartment. J Exp Med 181:2129–2140

Cascalho M, Ma A, Lee S, Masat L, Wabl M (1996) A quasi-monoclonal mouse. Science 272:1649–1652

Casellas R, Shih TA, Kleinewietfeld M, Rakonjac J, Nemazee D, Rajewsky K, Nussenzweig MC (2001) Contribution of receptor editing to the antibody repertoire. Science 291:1541–1544

Chen C, Nagy Z, Prak EL, Weigert M (1995) Immunoglobulin heavy chain gene replacement: a mechanism of receptor editing. Immunity 3:747–755

Chen C, Prak EL, Weigert M (1997) Editing disease-associated autoantibodies. Immunity 6:97–105

Chen Y, Zhang J, Hwang KK, Bouton-Verville H, Xia SM, Newman A, Ouyang YB, Haynes BF, Verkoczy L (2013) Common tolerance mechanisms, but distinct cross-reactivities associated with gp41 and lipids, limit production of HIV-1 broad neutralizing antibodies 2F5 and 4E10. J Immunol

Chumley MJ, Dal Porto JM, Kawaguchi S, Cambier JC, Nemazee D, Hardy RR (2000) A VH11 V kappa 9 B cell antigen receptor drives generation of CD5+ B cells both in vivo and in vitro. J Immunol 164:4586–4593

Clarke SH, Staudt LM, Kavaler J, Schwartz D, Gerhard WU, Weigert MG (1990) V region gene usage and somatic mutation in the primary and secondary responses to influenza virus hemagglutinin. J Immunol 144:2795–2801

Crews S, Griffin J, Huang H, Calame K, Hood L (1981) A single VH gene segment encodes the immune response to phosphorylcholine: somatic mutation is correlated with the class of the antibody. Cell 25:59–66

Doores KJ, Huber M, Le KM, Wang SK, Doyle-Cooper C, Cooper A, Pantophlet R, Wong CH, Nemazee D, Burton DR (2013) 2G12-expressing B cell lines may aid in HIV carbohydrate vaccine design strategies. J Virol 87:2234–2241

Dosenovic P, von Boehmer L, Escolano A, Jardine J, Freund NT, Gitlin AD, McGuire AT, Kulp DW, Oliveira T, Scharf L, Pietzsch J, Gray MD, Cupo A, van Gils MJ, Yao KH, Liu C, Gazumyan A, Seaman MS, Bjorkman PJ, Sanders RW, Moore JP, Stamatatos L, Schief WR, Nussenzweig MC (2015) Immunization for HIV-1 broadly neutralizing antibodies in human Ig knockin mice. Cell 161:1505–1515

Dougan SK, Ogata S, Hu CC, Grotenbreg GM, Guillen E, Jaenisch R, Ploegh HL (2012) IgG1 + ovalbumin-specific B-cell transnuclear mice show class switch recombination in rare allelically included B cells. Proc Natl Acad Sci USA 109:13739–13744

Doyle-Cooper C, Hudson KE, Cooper AB, Ota T, Skog P, Dawson PE, Zwick MB, Schief WR, Burton DR, Nemazee D (2013) Immune tolerance negatively regulates B cells in knock-in mice expressing broadly neutralizing HIV antibody 4E10. J Immunol

Dreyfus C, Laursen NS, Kwaks T, Zuijdgeest D, Khayat R, Ekiert DC, Lee JH, Metlagel Z, Bujny MV, Jongeneelen M, van der Vlugt R, Lamrani M, Korse HJ, Geelen E, Sahin O, Sieuwerts M, Brakenhoff JP, Vogels R, Li OT, Poon LL, Peiris M, Koudstaal W, Ward AB, Wilson IA, Goudsmit J, Friesen RH (2012) Highly conserved protective epitopes on influenza B viruses. Science 337:1343–1348

Ekiert DC, Friesen RH, Bhabha G, Kwaks T, Jongeneelen M, Yu W, Ophorst C, Cox F, Korse HJ, Brandenburg B, Vogels R, Brakenhoff JP, Kompier R, Koldijk MH, Cornelissen LA, Poon LL, Peiris M, Koudstaal W, Wilson IA, Goudsmit J (2011) A highly conserved neutralizing epitope on group 2 influenza A viruses. Science 333:843–850

Erikson J, Radic MZ, Camper SA, Hardy RR, Carmack C, Weigert M (1991) Expression of anti-DNA immunoglobulin transgenes in non-autoimmune mice. Nature 349:331–334

Feeney AJ, Atkinson MJ, Cowan MJ, Escuro G, Lugo G (1996) A defective Vkappa A2 allele in Navajos which may play a role in increased susceptibility to haemophilus influenzae type b disease. J Clin Invest 97:2277–2282

Finton KA, Larimore K, Larman HB, Friend D, Correnti C, Rupert PB, Elledge SJ, Greenberg PD, Strong RK (2013) Autoreactivity and exceptional CDR plasticity (but not unusual polyspecificity) hinder elicitation of the anti-HIV antibody 4E10. PLoS Pathog 9:e1003639

Foster MH, Liu Q, Chen H, Nemazee D, Cooperstone BG (1997) Anti-laminin reactivity and glomerular immune deposition by in vitro recombinant antibodies. Autoimmunity 26:231–243

Fulcher DA, Basten A (1994) Reduced life span of anergic self-reactive B cells in a double-transgenic model. J Exp Med 179:125–134

Gaeta BA, Malming HR, Jackson KJ, Bain ME, Wilson P, Collins AM (2007) iHMMune-align: hidden Markov model-based alignment and identification of germline genes in rearranged immunoglobulin gene sequences. Bioinformatics 23:1580–1587

Gay D, Saunders T, Camper S, Weigert M (1993) Receptor editing: an approach by autoreactive B cells to escape tolerance. J Exp Med 177:999–1008

Gearhart PJ, Johnson ND, Douglas R, Hood L (1981) IgG antibodies to phosphorylcholine exhibit more diversity than their IgM counterparts. Nature 291:29–34

Gerdes T, Wabl M (2004) Autoreactivity and allelic inclusion in a B cell nuclear transfer mouse. Nat Immunol 5:1282–1287

Golub R, Martin D, Bertrand FE, Cascalho M, Wabl M, Wu GE (2001) VH gene replacement in thymocytes. J Immunol 166:855–860

Goodnow CC, Crosbie J, Adelstein S, Lavoie TB, Smith-Gill SJ, Brink RA, Pritchard-Briscoe H, Wotherspoon JS, Loblay RH, Raphael K et al. (1988) Altered immunoglobulin expression and functional silencing of self-reactive B lymphocytes in transgenic mice. Nature 334, 676–682

Goodnow CC, Sprent J, Fazekas de St GB, Vinuesa CG (2005) Cellular and genetic mechanisms of self tolerance and autoimmunity. Nature 435:590–597

Green LL (2014) Transgenic mouse strains as platforms for the successful discovery and development of human therapeutic monoclonal antibodies. Curr Drug Discov Technol 11:74–84

Hagman J, Lo D, Doglio LT, Hackett JJ, Rudin CM, Haasch D, Brinster R, Storb U (1989) Inhibition of immunoglobulin gene rearrangement by the expression of a lambda 2 transgene [published erratum appears in J Exp Med 1989 Aug 1;170(2):619]. J Exp Med 169, 1911–1929

Hangartner L, Senn BM, Ledermann B, Kalinke U, Seiler P, Bucher E, Zellweger RM, Fink K, Odermatt B, Burki K, Zinkernagel RM, Hengartner H (2003) Antiviral immune responses in gene-targeted mice expressing the immunoglobulin heavy chain of virus-neutralizing antibodies. Proc Natl Acad Sci USA 100:12883–12888

Hangartner L, Zellweger RM, Giobbi M, Weber J, Eschli B, McCoy KD, Harris N, Recher M, Zinkernagel RM, Hengartner H (2006) Nonneutralizing antibodies binding to the surface glycoprotein of lymphocytic choriomeningitis virus reduce early virus spread. J Exp Med 203:2033–2042

Hara T, Jung LK, Bjorndahl JM, Fu SM (1986) Human T cell activation. III. Rapid induction of a phosphorylated 28 kD/32 kD disulfide-linked early activation antigen (EA 1) by 12-o-tetradecanoyl phorbol-13-acetate, mitogens, and antigens. J Exp Med 164:1988–2005

Hayakawa K, Asano M, Shinton SA, Gui M, Allman D, Stewart CL, Silver J, Hardy RR (1999) Positive selection of natural autoreactive B cells. Science 285:113–116

Hayakawa K, Asano M, Shinton SA, Gui M, Wen LJ, Dashoff J, Hardy RR (2003) Positive selection of anti-thy-1 autoreactive B-1 cells and natural serum autoantibody production independent from bone marrow B cell development. J Exp Med 197:87–99

Haynes BF, Fleming J, St Clair EW, Katinger H, Stiegler G, Kunert R, Robinson J, Scearce RM, Plonk K, Staats HF, Ortel TL, Liao HX, Alam SM (2005) Cardiolipin polyspecific autoreactivity in two broadly neutralizing HIV-1 antibodies. Science 308:1906–1908

Hochedlinger K, Jaenisch R (2002) Monoclonal mice generated by nuclear transfer from mature B and T donor cells. Nature 415:1035–1038

Hombach J, Tsubata T, Leclercq L, Stappert H, Reth M (1990) Molecular components of the B-cell antigen receptor complex of the IgM class. Nature 343:760–762

Hoot S, McGuire AT, Cohen KW, Strong RK, Hangartner L, Klein F, Diskin R, Scheid JF, Sather DN, Burton DR, Stamatatos L (2013) Recombinant HIV envelope proteins fail to engage germline versions of anti-CD4bs bNAbs. PLoS Pathog 9:e1003106

Hu L, Rezanka LJ, Mi QS, Lustig A, Taub DD, Longo DL, Kenny JJ (2002) T15-idiotype-negative B cells dominate the phosphocholine binding cells in the preimmune repertoire of T15i knockin mice. J Immunol 168:1273–1280

Hur EM, Patel SN, Shimizu S, Rao DS, Gnanapragasam PN, An DS, Yang L, Baltimore D (2012) Inhibitory effect of HIV-specific neutralizing IgA on mucosal transmission of HIV in humanized mice. Blood 120:4571–4582

Jardine J, Julien JP, Menis S, Ota T, Kalyuzhniy O, McGuire A, Sok D, Huang PS, Macpherson S, Jones M, Nieusma T, Mathison J, Baker D, Ward AB, Burton DR, Stamatatos L, Nemazee D, Wilson IA, Schief WR (2013) Rational HIV immunogen design to target specific germline B cell receptors. Science 340:711–716

Jardine JG, Ota T, Sok D, Pauthner M, Kulp DW, Kalyuzhniy O, Skog PD, Thinnes TC, Bhullar D, Briney B, Menis S, Jones M, Kubitz M, Spencer S, Adachi Y, Burton DR, Schief WR, Nemazee D (2015) HIV-1 VACCINES. Priming a broadly neutralizing antibody response to HIV-1 using a germline-targeting immunogen. Science 349:156–161

Kaartinen M, Griffiths GM, Markham AF, Milstein C (1983) mRNA sequences define an unusually restricted IgG response to 2-phenyloxazolone and its early diversification. Nature 304:320–324

Karpel ME, Boutwell CL, Allen TM (2015) BLT humanized mice as a small animal model of HIV infection. Curr Opin Virol 13:75–80

Kashyap AK, Steel J, Oner AF, Dillon MA, Swale RE, Wall KM, Perry KJ, Faynboym A, Ilhan M, Horowitz M, Horowitz L, Palese P, Bhatt RR, Lerner RA (2008) Combinatorial antibody libraries from survivors of the Turkish H5N1 avian influenza outbreak reveal virus neutralization strategies. Proc Natl Acad Sci USA 105:5986–5991

Kenny JJ, Stall AM, Sieckmann DG, Lamers MC, Finkelman FD, Finch L, Longo DL (1991) Receptor-mediated elimination of phosphocholine-specific B cells in x-linked immune-deficient mice. J Immunol 146:2568–2577

Kirak O, Frickel EM, Grotenbreg GM, Suh H, Jaenisch R, Ploegh HL (2010) Transnuclear mice with predefined T cell receptor specificities against Toxoplasma gondii obtained via SCNT. Science 328:243–248

Kitamura D, Rajewsky K (1992) Targeted disruption of mu chain membrane exon causes loss of heavy-chain allelic exclusion [see comments]. Nature 356:154–156

Klein F, Halper-Stromberg A, Horwitz JA, Gruell H, Scheid JF, Bournazos S, Mouquet H, Spatz LA, Diskin R, Abadir A, Zang T, Dorner M, Billerbeck E, Labitt RN, Gaebler C, Marcovecchio PM, Incesu RB, Eisenreich TR, Bieniasz PD, Seaman MS, Bjorkman PJ, Ravetch JV, Ploss A, Nussenzweig MC (2012) HIV therapy by a combination of broadly neutralizing antibodies in humanized mice. Nature 492:118–122

Koralov SB, Novobrantseva TI, Konigsmann J, Ehlich A, Rajewsky K (2006) Antibody repertoires generated by VH replacement and direct VH to JH joining. Immunity 25:43–53

Kumar R, Bach MP, Mainoldi F, Maruya M, Kishigami S, Jumaa H, Wakayama T, Kanagawa O, Fagarasan S, Casola S (2015) Antibody repertoire diversification through VH gene replacement in mice cloned from an IgA plasma cell. Proc Natl Acad Sci USA 112:E450–457

Lenschow DJ, Sperling AI, Cooke MP, Freeman G, Rhee L, Decker DC, Gray G, Nadler LM, Goodnow CC, Bluestone JA (1994) Differential up-regulation of the B7-1 and B7-2 costimulatory molecules after Ig receptor engagement by antigen. J Immunol 153:1990–1997

Lieberman R, Potter M, Mushinski EB, Humphrey W Jr, Rudikoff S (1974) Genetics of a new IgVH (T15 idiotype) marker in the mouse regulating natural antibody to phosphorylcholine. J Exp Med 139:983–1001

Lingwood D, McTamney PM, Yassine HM, Whittle JR, Guo X, Boyington JC, Wei CJ, Nabel GJ (2012) Structural and genetic basis for development of broadly neutralizing influenza antibodies. Nature

Litzenburger T, Bluthmann H, Morales P, Pham-Dinh D, Dautigny A, Wekerle H, Iglesias A (2000) Development of myelin oligodendrocyte glycoprotein autoreactive transgenic B lymphocytes: receptor editing in vivo after encounter of a self-antigen distinct from myelin oligodendrocyte glycoprotein. J Immunol 165:5360–5366

Lucas AH, Reason DC (1999) Polysaccharide vaccines as probes of antibody repertoires in man. Immunol Rev 171:89–104

Luning Prak E, Weigert M (1995) Light chain replacement: a new model for antibody gene rearrangement. J Exp Med 182:541–548

Ma B, Osborn MJ, Avis S, Ouisse LH, Menoret S, Anegon I, Buelow R, Bruggemann M (2013) Human antibody expression in transgenic rats: comparison of chimeric IgH loci with human VH, D and JH but bearing different rat C-gene regions. J Immunol Methods 400–401:78–86

Martin F, Oliver AM, Kearney JF (2001) Marginal zone and B1 B cells unite in the early response against T-independent blood-borne particulate antigens. Immunity 14:617–629

McGuire AT, Hoot S, Dreyer AM, Lippy A, Stuart A, Cohen KW, Jardine J, Menis S, Scheid JF, West AP, Schief WR, Stamatatos L (2013) Engineering HIV envelope protein to activate germline B cell receptors of broadly neutralizing anti-CD4 binding site antibodies. J Exp Med 210:655–663

McGuire AT, Dreyer AM, Carbonetti S, Lippy A, Glenn J, Scheid JF, Mouquet H, Stamatatos L (2014a) HIV antibodies. Antigen modification regulates competition of broad and narrow neutralizing HIV antibodies. Science 346:1380–1383

McGuire AT, Glenn JA, Lippy A, Stamatatos L (2014b) Diverse recombinant HIV-1 Envs fail to activate B cells expressing the germline B cell receptors of the broadly neutralizing anti-HIV-1 antibodies PG9 and 447-52D. J Virol 88:2645–2657

Mi QS, Zhou L, Schulze DH, Fischer RT, Lustig A, Rezanka LJ, Donovan DM, Longo DL, Kenny JJ (2000) Highly reduced protection against Streptococcus pneumoniae after deletion of a single heavy chain gene in mouse. Proc Natl Acad Sci USA 97:6031–6036

Nadel B, Tang A, Lugo G, Love V, Escuro G, Feeney AJ (1998) Decreased frequency of rearrangement due to the synergistic effect of nucleotide changes in the heptamer and nonamer of the recombination signal sequence of the V kappa gene A2b, which is associated with increased susceptibility of Navajos to *Haemophilus influenzae* type b disease. J Immunol 161:6068–6073

Nemazee D (2006) Receptor editing in lymphocyte development and central tolerance. Nat Rev Immunol 6:728–740

Nemazee DA, Burki K (1989) Clonal deletion of B lymphocytes in a transgenic mouse bearing anti-MHC class I antibody genes. Nature 337:562–566

Nisonoff A, Ju S (1976) Studies of a cross-reactive idiotype associated with anti-para-azophenylarsonate antibodies of A/J mice. Ann Immunol (Paris) 127:347–356

Nussenzweig MC, Shaw AC, Sinn E, Danner DB, Holmes KL, Morse HC, Leder P (1987) Allelic exclusion in transgenic mice that express the membrane form of immunoglobulin mu. Science 236:816–819

O'Brien RL, Brinster RL, Storb U (1987) Somatic hypermutation of an immunoglobulin transgene in kappa transgenic mice. Nature 326:405–409

Osborn MJ, Ma B, Avis S, Binnie A, Dilley J, Yang X, Lindquist K, Menoret S, Iscache AL, Ouisse LH, Rajpal A, Anegon I, Neuberger MS, Buelow R, Bruggemann M (2013) High-affinity IgG antibodies develop naturally in Ig-knockout rats carrying germline human IgH/Igkappa/Iglambda loci bearing the rat CH region. J Immunol 190:1481–1490

Ota T, Doyle-Cooper C, Cooper AB, Huber M, Falkowska E, Doores KJ, Hangartner L, Le K, Sok D, Jardine J, Lifson J, Wu X, Mascola JR, Poignard P, Binley JM, Chakrabarti BK, Schief WR, Wyatt RT, Burton DR, Nemazee D (2012) Anti-HIV B cell lines as candidate vaccine biosensors. J. Immunol. 189:4816–4824

Ota T, Doyle-Cooper C, Cooper AB, Doores KJ, Aoki-Ota M, Le K, Schief WR, Wyatt RT, Burton DR, Nemazee D (2013) B cells from knock-in mice expressing broadly neutralizing HIV antibody b12 carry an innocuous B cell receptor responsive to HIV vaccine candidates. J Immunol

Parameswaran P, Liu Y, Roskin KM, Jackson KK, Dixit VP, Lee JY, Artiles KL, Zompi S, Vargas MJ, Simen BB, Hanczaruk B, McGowan KR, Tariq MA, Pourmand N, Koller D, Balmaseda A, Boyd SD, Harris E, Fire AZ (2013) Convergent antibody signatures in human dengue. Cell Host Microbe 13:691–700

Pelanda R, Schaal S, Torres RM, Rajewsky K (1996) A prematurely expressed Ig(kappa) transgene, but not V(kappa)J(kappa) gene segment targeted into the Ig(kappa) locus, can rescue B cell development in lambda5-deficient mice. Immunity 5:229–239

Pelanda R, Schwers S, Sonoda E, Torres RM, Nemazee D, Rajewsky K (1997) Receptor editing in a transgenic mouse model: site, efficiency, and role in B cell tolerance and antibody diversification. Immunity 7:765–775

Pewzner-Jung Y, Friedmann D, Sonoda E, Jung S, Rajewsky K, Eilat D (1998) B cell deletion, anergy, and receptor editing in "knock in" mice targeted with a germline-encoded or somatically mutated anti-DNA heavy chain. J Immunol 161:4634–4645

Phan TG, Amesbury M, Gardam S, Crosbie J, Hasbold J, Hodgkin PD, Basten A, Brink R (2003) B cell receptor-independent stimuli trigger immunoglobulin (Ig) class switch recombination and production of IgG autoantibodies by anergic self-reactive B cells. J Exp Med 197:845–860

Pruzina S, Williams GT, Kaneva G, Davies SL, Martin-Lopez A, Bruggemann M, Vieira SM, Jeffs SA, Sattentau QJ, Neuberger MS (2011) Human monoclonal antibodies to HIV-1 gp140 from mice bearing YAC-based human immunoglobulin transloci. Protein Eng Des Sel 24:791–799

Reth M (1989) Antigen receptor tail clue. Nature 338:383–384

Reth M (1992) Antigen receptors on B lymphocytes. Annu Rev Immunol 10:97–121

Ritchie KA, Brinster RL, Storb U (1984) Allelic exclusion and control of endogenous immunoglobulin gene rearrangement in kappa transgenic mice. Nature 312:517–520

Rusconi S, Kohler G (1985) Transmission and expression of a specific pair of rearranged immunoglobulin mu and kappa genes in a transgenic mouse line. Nature 314:330–334

Russ DE, Ho KY, Longo NS (2015) HTJoinSolver: human immunoglobulin VDJ partitioning using approximate dynamic programming constrained by conserved motifs. BMC Bioinformatics 16:170

Russell DM, Dembic Z, Morahan G, Miller JF, Burki K, Nemazee D (1991) Peripheral deletion of self-reactive B cells. Nature 354:308–311

Seiler P, Kalinke U, Rulicke T, Bucher EM, Bose C, Zinkernagel RM, Hengartner H (1998) Enhanced virus clearance by early inducible lymphocytic choriomeningitis virus-neutralizing antibodies in immunoglobulin-transgenic mice. J Virol 72:2253–2258

Senn BM, Lopez-Macias C, Kalinke U, Lamarre A, Isibasi A, Zinkernagel RM, Hengartner H (2003) Combinatorial immunoglobulin light chain variability creates sufficient B cell diversity to mount protective antibody responses against pathogen infections. Eur J Immunol 33:950–961

Sher A, Cohn M (1972) Inheritance of an idiotype associated with the immune response of inbred mice to phosphorylcholine. Eur J Immunol 2:319–326

Shlomchik MJ (2008) Sites and stages of autoreactive B cell activation and regulation. Immunity 28:18–28

Shlomchik MJ, Zharhary D, Saunders T, Camper SA, Weigert Mg (1993) A rheumatoid factor transgenic mouse model of autoantibody regulation. Int Immunol 5, 1329–1341

Sigal NH, Pickard AR, Metcalf ES, Gearhart PJ, Klinman NR (1977) Expression of phosphorylcholine-specific B cells during murine development. J Exp Med 146:933–948

Sonoda E, Pewzner-Jung Y, Schwers S, Taki S, Jung S, Eilat D, Rajewsky K (1997) B cell development under the condition of allelic inclusion. Immunity 6:225–233

Storb U (1987) Transgenic mice with immunoglobulin genes. Annu Rev Immunol 5:151–174

Storb U, Pinkert C, Arp B, Engler P, Gollahon K, Manz J, Brady W, Brinster RL (1986) Transgenic mice with mu and kappa genes encoding antiphosphorylcholine antibodies. J Exp Med 164:627–641

Sui J, Hwang WC, Perez S, Wei G, Aird D, Chen LM, Santelli E, Stec B, Cadwell G, Ali M, Wan H, Murakami A, Yammanuru A, Han T, Cox NJ, Bankston LA, Donis RO, Liddington RC, Marasco WA (2009) Structural and functional bases for broad-spectrum neutralization of avian and human influenza A viruses. Nat Struct Mol Biol 16:265–273

Taki S, Meiering M, Rajewsky K (1993) Targeted insertion of a variable region gene into the immunoglobulin heavy chain locus [see comments]. Science 262:1268–1271

Taki S, Schwenk F, Rajewsky K (1995) Rearrangement of upstream DH and VH genes to a rearranged immunoglobulin variable region gene inserted into the DQ52-JH region of the immunoglobulin heavy chain locus. Eur J Immunol 25:1888–1896

Thomson CA, Bryson S, McLean GR, Creagh AL, Pai EF, Schrader JW (2008) Germline V-genes sculpt the binding site of a family of antibodies neutralizing human cytomegalovirus. EMBO J 27:2592–2602

Throsby M, van den Brink E, Jongeneelen M, Poon LL, Alard P, Cornelissen L, Bakker A, Cox F, van Deventer E, Guan Y, Cinatl J, ter Meulen J, Lasters I, Carsetti R, Peiris M, de Kruif J, Goudsmit J (2008) Heterosubtypic neutralizing monoclonal antibodies cross-protective against H5N1 and H1N1 recovered from human IgM + memory B cells. PLoS ONE 3:e3942

Tiegs SL, Russell DM, Nemazee D (1993) Receptor editing in self-reactive bone marrow B cells. J Exp Med 177:1009–1020

Venkitaraman AR, Williams GT, Dariavach P, Neuberger MS (1991) The B-cell antigen receptor of the five immunoglobulin classes. Nature 352:777–781

Verkoczy L, Diaz M (2014) Autoreactivity in HIV-1 broadly neutralizing antibodies: implications for their function and induction by vaccination. Curr Opin HIV and AIDS 9:224–234

Verkoczy L, Diaz M, Holl TM, Ouyang YB, Bouton-Verville H, Alam SM, Liao HX, Kelsoe G, Haynes BF (2010) Autoreactivity in an HIV-1 broadly reactive neutralizing antibody variable region heavy chain induces immunologic tolerance. Proc Natl Acad Sci USA 107:181–186

Verkoczy L, Chen Y, Bouton-Verville H, Zhang J, Diaz M, Hutchinson J, Ouyang YB, Alam SM, Holl TM, Hwang KK, Kelsoe G, Haynes BF (2011) Rescue of HIV-1 broad neutralizing antibody-expressing B cells in 2F5 VH x VL knockin mice reveals multiple tolerance controls. J Immunol 187:3785–3797

Verkoczy L, Chen Y, Zhang J, Bouton-Verville H, Newman A, Lockwood B, Scearce RM, Montefiori DC, Dennison SM, Xia SM, Hwang KK, Liao HX, Alam SM, Haynes BF (2013) Induction of HIV-1 broad neutralizing antibodies in 2F5 knock-in mice: selection against membrane proximal external region-associated autoreactivity limits T-dependent responses. J Immunol 191:2538–2550

Villaudy J, Schotte R, Legrand N, Spits H (2014) Critical assessment of human antibody generation in humanized mouse models. J Immunol Methods 410:18–27

Weaver D, Costantini F, Imanishi-Kari T, Baltimore D (1985) A transgenic immunoglobulin mu gene prevents rearrangement of endogenous genes. Cell 42:117–127

Weaver GC, Villar RF, Kanekiyo M, Nabel GJ, Mascola JR, Lingwood D (2016) In vitro reconstitution of B cell receptor-antigen interactions to evaluate potential vaccine candidates. Nat Protoc 11:193–213

Whittle JR, Zhang R, Khurana S, King LR, Manischewitz J, Golding H, Dormitzer PR, Haynes BF, Walter EB, Moody MA, Kepler TB, Liao HX, Harrison SC (2011) Broadly neutralizing human antibody that recognizes the receptor-binding pocket of influenza virus hemagglutinin. Proc Natl Acad Sci USA 108:14216–14221

Whittle JR, Wheatley AK, Wu L, Lingwood D, Kanekiyo M, Ma SS, Narpala SR, Yassine HM, Frank GM, Yewdell JW, Ledgerwood JE, Wei CJ, McDermott AB, Graham BS, Koup RA, Nabel GJ (2014) Flow cytometry reveals that H5N1 vaccination elicits cross-reactive stem-directed antibodies from multiple Ig heavy-chain lineages. J Virol 88:4047–4057

Wrammert J, Koutsonanos D, Li GM, Edupuganti S, Sui J, Morrissey M, McCausland M, Skountzou I, Hornig M, Lipkin WI, Mehta A, Razavi B, Del Rio C, Zheng NY, Lee JH, Huang M, Ali Z, Kaur K, Andrews S, Amara RR, Wang Y, Das SR, O'Donnell CD, Yewdell JW, Subbarao K, Marasco WA, Mulligan MJ, Compans R, Ahmed R, Wilson PC (2011) Broadly cross-reactive antibodies dominate the human B cell response against 2009 pandemic H1N1 influenza virus infection. J Exp Med 208:181–193

Wu X, Zhou T, Zhu J, Zhang B, Georgiev I, Wang C, Chen X, Longo NS, Louder M, McKee K, O'Dell S, Perfetto S, Schmidt SD, Shi W, Wu L, Yang Y, Yang ZY, Yang Z, Zhang Z, Bonsignori M, Crump JA, Kapiga SH, Sam NE, Haynes BF, Simek M, Burton DR, Koff WC, Doria-Rose NA, Connors M, Mullikin JC, Nabel GJ, Roederer M, Shapiro L, Kwong PD, Mascola JR (2011) Focused evolution of HIV-1 neutralizing antibodies revealed by structures and deep sequencing. Science 333:1593–1602

Wysocki LJ, Margolies MN, Huang B, Nemazee DA, Wechsler DS, Sato VL, Smith JA, Gefter ML (1985) Combinational diversity within variable regions bearing the predominant anti-p-azophenylarsonate idiotype of strain A mice. J Immunol 134:2740–2747

Yang G, Holl TM, Liu Y, Li Y, Lu X, Nicely NI, Kepler TB, Alam SM, Liao HX, Cain DW, Spicer L, Vandeberg JL, Haynes BF, Kelsoe G (2013) Identification of autoantigens recognized by the 2F5 and 4E10 broadly neutralizing HIV-1 antibodies. J Exp, Med

Ye J, Ma N, Madden TL, Ostell JM (2013) IgBLAST: an immunoglobulin variable domain sequence analysis tool. Nucleic Acids Res 41:W34–40

Zhang R, Verkoczy L, Wiehe K, Munir Alam S, Nicely NI, Santra S, Bradley T, Pemble CWt., Zhang J, Gao F, Montefiori DC, Bouton-Verville H, Kelsoe G, Larimore K, Greenberg PD, Parks R, Foulger A, Peel JN, Luo K, Lu X, Trama AM, Vandergrift N, Tomaras GD,

Kepler TB, Moody MA, Liao HX, Haynes BF (2016) Initiation of immune tolerance-controlled HIV gp41 neutralizing B cell lineages. Sci Transl Med 8, 336ra362

Zhou J, Lottenbach KR, Barenkamp SJ, Lucas AH, Reason DC (2002) Recurrent variable region gene usage and somatic mutation in the human antibody response to the capsular polysaccharide of *Streptococcus pneumoniae* type 23F. Infect Immun 70:4083–4091

Zhou T, Georgiev I, Wu X, Yang ZY, Dai K, Finzi A, Kwon YD, Scheid JF, Shi W, Xu L, Yang Y, Zhu J, Nussenzweig MC, Sodroski J, Shapiro L, Nabel GJ, Mascola JR, Kwong PD (2010) Structural basis for broad and potent neutralization of HIV-1 by antibody VRC01. Science 329:811–817

Zhou T, Zhu J, Wu X, Moquin S, Zhang B, Acharya P, Georgiev IS, Altae-Tran HR, Chuang GY, Joyce MG, Do Kwon Y, Longo NS, Louder MK, Luongo T, McKee K, Schramm CA, Skinner J, Yang Y, Yang Z, Zhang Z, Zheng A, Bonsignori M, Haynes BF, Scheid JF, Nussenzweig MC, Simek M, Burton DR, Koff WC, Program NCS, Mullikin JC, Connors M, Shapiro L, Nabel GJ, Mascola JR, Kwong PD (2013) Multidonor analysis reveals structural elements, genetic determinants, and maturation pathway for HIV-1 neutralization by VRC01-class antibodies. Immunity 39:245–258

CPSIA information can be obtained
at www.ICGtesting.com
Printed in the USA
LVHW082147080920
665396LV00002B/3